NORTH CAROLINA

23RD REGIMENT INFANTRY

ROSTER AND HISTORY

John W. Moore
and
Walter Clark

Heritage Books
2026

HERITAGE BOOKS

AN IMPRINT OF HERITAGE BOOKS, INC.

Books, CDs, and more—Worldwide

For our listing of thousands of titles see our website
at
www.HeritageBooks.com

A Facsimile Reprint
Published 2026 by
HERITAGE BOOKS, INC.
Publishing Division
5810 Ruatan Street
Berwyn Heights, MD 20740

Reprinted 2019
Mountain Press
Signal Mountain, Tennessee

International Standard Book Number
Paperbound: 978-0-7884-9248-8

NORTH CAROLINA
23rd REGIMENT INFANTRY
ROSTER and HISTORY

This volume of the Regimental Infantry contains both the Roster and the History of the unit. Each section of the material has somewhat different information and additional data on each of the soldiers in this unit.

The Roster section of this volume contains a listing of each of the men or women in the Companies. In the Roster portion of this volume you will find a number of items of interest on each of the men. You can find, using the following guide: c = captured - cm = commissioned - co = county - COM = Company - e = enlisted - d = died - dg = discharged - dt = detailed - k = killed - m = missing - p = promoted - pr = prisoner - r = resigned - tr = transferred and w = wounded.

In the History portion there is a listing of each of the Companies and the story of the action of the Companies during their time in battle and between battles. There are two different portions written by two different men, i.e. Vines E. Turner and H. Clay Wall. Each of the two versions tell the same story, but from different viewpoints thus giving a more full history of this unit in their action.

Company A - Capt. Wm. F. Harlee - the Anson Ellis Rifles
Company B - Capt. George W. Seagle - the Hog Hill Guards - Lincoln County
Company C - Capt. C. J. Cochrane - the Montgomery Volunteers No. 1
Company D - Capt. Lewis H. Webb - the Pee Dee Guards - Richmond County
Company E - Capt. J. H. Horner - Granville Plough Boys
Company F - Capt. M. L. McCorkle - Catawba Guards
Company G.- Capt. C. C. Blacknall - Granville Rifles
Company H - Capt. E. M. Faires - the Gaston Guards
Company I - Capt. Rufus Amis - Granville Stars
Company K - Capt. Robert D. Johnston - Beattie's Ford Riflemen - Lincoln County

In doing research in this volume, study the Roster first and then carefully read the History to get a full picture of the Regiment during the whole of the Civil War. These Confederate units suffered heavy losses in most of the battles in which they fought and if a soldier survived until the end of the war, they were very lucky. Many of these young men left wives and children behind in order to fight and many did not return home.

James L. Douthat
Mountain Press
2019

TWENTY-THIRD REGIMENT.

1. R. D. Johnston, Colonel.   4. C. C. Blacknall, Colonel.
2. J. F. Hoke, Colonel.   5. J. W. Leak, Lieut.-Colonel.
3. D. H. Christie, Colonel.   6. E. J. Christian, Major.
7. Rev. Theophilus W. Moore, Chaplain.

TWENTY-THIRD REGIMENT—INFANTRY.

ABBREVIATIONS.

c....................captured.	d.......................died.	p.....................promoted.
cm.............commissioned.	dg. ....discharged.	pr.......................prisoner.
co...........................county.	dt....................detailed.	r........................resigned.
Com..................Company.	k.................... killed.	tr.................. transferred.
e............................enlisted.	m..................missing.	w....................wounded.

FIELD AND STAFF.

John F. Hoke, Colonel, cm. June 10th, '61; Lincoln co.

Daniel H. Christie, Colonel, cm. May 10th, '62; Granville co.; w. July 1st, '63, at Gettysburg; d. in Winchester August, '63.

Charles C. Blacknall, Colonel, cm. Aug. 15th, '63; Granville co.; pr. at Gettysburg; w. at Winchester Sept. 19th, '64; d. Oct. 20th, '64.

Wm. S. Davis, Colonel, cm. Oct., '64; Warren co.; tr. from 12th Regiment; w.

John W. Leak, Lieut. Colonel, cm. June 10th, '61; Richmond co.

Robert D. Johnston, Lieut. Colonel, cm. May 10th, '62; Lincoln co.; w. at Seven Pines; p. Brigadier General July, '63; w.

Daniel H. Christie, Major, cm. June 10th, '61; Granville co.; p. from Captain of Com. G and k.

E. J. Christian, Major, cm. April 26th, '62; Montgomery co.; k. May 31st, '62, at Seven Pines.

Charles C. Blacknall, Major, cm. May 31st, '62; Granville co.; p. from Captain of Com. G.

Isaac Jones Young, Adjutant, cm. June 10th, '61; Granville co.; w. July 1st, '62: p. Captain of Com. G and r. in '62.

Vines E. Turner, Adjutant, cm. May 10th, '62; Granville co.; w. June 27th, '62.

Junius French, Adjutant, cm. June, '63; Yadkin co.; k. July 1st, '63, at Gettysburg.

Charles P. Powell, Adjutant, cm. July, '63; Richmond co.; k. May 12th, '64, at Spottsylvania C. H.

Lawrence Everett, Adjutant, cm. May 12th, '64; Richmond co.

Edwin G. Cheatham, A. Q. M., Granville co.; cm. June 10th, '61; r. Feb., '62.

W. I. Everett, A. Q. M., cm. in '62; Richmond co.; r.

Vines E. Turner, A. Q. M., cm. June, '63; Granville co.

James F. Johnston, A. C. S., Lincoln co.

Theophilus Moore, Chaplain, Person co.

Robert J. Hicks, Surgeon, Granville co.

William F. Gill, Sergeant-Major, Granville co.; k. July 1st, '62, at Malvern Hill.

COMPANY A.

OFFICERS.

William F. Harlee, Captain, cm. May 22d, '61; Anson co.; r. December 15th, '61.

James M. Wall, Captain, cm. December 15th, '61; Anson co.

Frank Bennett, Captain, cm. May 10th, '62; Anson co.; p. from 1st Sergeant; w.

W. D. Redfearne, 1st Lieut., cm. May 22d, '61; Anson co.

James C. Marshall, 1st Lieut., cm. May 10th, 62; Anson co.; tr. as Adjutant to 15th Regiment in '62.

John M. Little, 2d Lieut., cm. May 22d, '61; Anson co.

James Crowder, 2d Lieut., cm. May 22d, '61; Anson co.

Samuel F. Wright, 2d Lieut., cm. May 10th, '62; Anson co.

NON-COMMISSIONED OFFICERS.

Frank Bennett, 1st Sergeant, e. May 22d, '61; Anson co.; p. Captain May 10th, '62.

Clem. G. Richardson, 2d Sergeant, e. May 10th, '62; Anson co.; dg. December 1st, '61, for disability.

William A. Lowman, 3d Sergeant, e. May 22d, '61; Anson eo.; d. July 10th, '61, at Raleigh.

James M. Wall, 4th Sergeant, e. May 22d, '61; Anson co.; p. Captain Dec. 15th, '61.

Joseph Crowder, 5th Sergeant, e. May 22d, '61; Anson co.; pr. at Gettysburg.

Alexander Duren, 1st Corporal, e. May 22d, '61; Anson co.; p. Sergeant.

M. H. Allen, 2d Corporal, e. May 22d, '61; Anson co.; w. and pr. at Gettysburg.

Sam'l F. Wright, 3d Corporal, e. May 22d, '61; Anson co.; p. 2d Lieut. May 10th, '62.

W. W. Carpenter, 4th Corporal, e. May 22d, '61; Anson co.; d. September 20th, '61, at Gordonsville.

PRIVATES.

Allen, M. H., e. August 22d, '61; Anson co.

Allen, Robert M., e. August 22d, '61; Anson co.; p. Corporal and pr. September 19th, '64.

Barrett, Wm. A., e. August 22d, '61; Anson co.; d. August 4th, '62, in Va.

Barnett, Armstrong, e. August 22d, '61; Anson co.; k. May 3d, '63, at Chancellorsville.

Billingsby, Frank, e. August 22d, '61; Anson co.; k. May 3d, '63, at Chancellorsville.

Billingsby, J. D., e. August 22d, '61; Anson co.; k. May 3d, '63, at Chancellorsville.

Burmingham, Wm., e. August 22d, '61; Anson co.; w. at Seven Pines, Sharpsburg and Gettysburg.

Blackwell, Stephen W., e. August 22d, '61; Anson co.; dg. Dec. 18th, '61.

Bowers, Frank, e. August 22d, '61; Anson co.; d. of w. received at Gettysburg.

Broadway, W. M., e. August 22d, '61; Anson co.; dg. Jan. 2d, '62.

Bullard, B. T., e. August 22d, '61; Anson co.; dg. January 2nd, '62.

Beverly, W. P., e. May 15th, '61; Anson co.

Billingsby, J. M., e. Feb. 28th, '63; Anson co.; w. and pr. at Gettysburg.

Barger, Hosea, e. Sept. 6th, '62; Caawba co.; w. at Chancellorsville.

Biggers, R. W., e. October 15th, '64; Wake co.

Campbell, John, e. Sept. 6th, '62; Catawba co.; m.

Crowder, Richard, e. May 15th, '62; Anson co.; w. at Sharpsburg and Gettysburg.

Courtney, C. C., e. May 22d, '61; Anson co; k. July 1st, '62, at Malvern Hill.

Clark, W. C., e. August 22d, '61; Anson co.; dg. Oct. 10th, '61.

Cenney, L. C., e. August 22d, '61; Anson co.; dg. Oct. 1st, '62.

Cox, Jesse J., e. August 22d, '61; Anson co.; w. and pr. at Chancellorsville.

Crump, James F., e. August 22d, '61; Anson co.

Dennis, Charles J. F., e. August 22d, '61; Anson co.; pr. at Gettysburg.

Dunlap, William P., e. August 22d, '61; Anson co.; k. May 3d, '63, at Chancellorsville.

Drum, W. H., e. Sept. 6th, '62; Alexander co.

Eades, J. N., e. Sept. 6th, '62; Catawba co.; pr. at Gettysburg.

Eckards, Wm., e. Sept. 6th, '62; Catawba co.

Eddleman, H. M., e. Sept. 6th, '62; Catawba co.; m.

Earnhardt, Peter S., e. August 22d, '61; Anson co.; p. Sergeant and w. at Gettysburg.

Farmer, S., e. August 22d, '61; Anson co.

Freeman, M. C., e. ——; Wake co.

Farmer, Seabury, e. May 22d, '61; Anson co.; d. Nov. 30th, '61, at Manassas.

Grissom, Wm., e. Dec. 5th, '62; Anson co.; d. Jan. 15th, '63, at Lynchburg.

Gaddy, John, e. August 22d, '61; Anson co.; w. Oct. 12th, '63, at Stevensburg.

Gaddy, Franklin, e. August 22d, '61; Anson co.; d. Sept. 12th, '61, at Manassas.

Gaddy, Dennis D., e. August 22d, '61; Anson co.; w. at Chancellorsville and pr. at Gettysburg.

Gaddy, Patrick P., e. August 22d, '61; Anson co.; w. at Sharpsburg, and k. May 3d, '63, at Chancellorsville.

Gatewood, Philip, e. Aug. 22d, '61; Anson co.; d. Aug. 25th, '61, at Richmond.

Gardner, N. H., e. Aug. 22d, '61; Anson co.; d. June 11th, '63, at Williamsport.

Grady, John, e. Aug. 22d, '61; Anson co.; w. and pr. at Gettysburg.

Guledge, Joel, e. Aug. 22d, '61; Anson co.; pr. at Gettysburg.

Horn, John, e. Aug. 22d, '61; Anson co.

Horn, Jas. T., e. Aug. 22d, '61; Anson co.; dg. Sept. 31st, '62.

Horn, Martin V., e. Aug. 22d, '61; Anson co.; dg. Sept. 3d, '62.

Horn, Wm., e. Aug. 22d, '61; Anson co.; d. May 5th, '63, of w. at Chancellorsville.

Hendrick, John D., e. Aug. 22d, '61; Anson co.; w. at Chancellorsville.

Hoss, R. M., e. Sept. 6th, '62; Catawba co.; pr. at Gettysburg.

Henman, Peter, e. Sept. 6th, '62; Catawba co.; d.

Hevener, Peter, e. Sept. 6th, '62; Lincoln co.

Hayse, Jackson, e. Sept. 6th, '62; Catawba co.

Horn, Lewis I., e. May 22d, '61; Anson co.; d. May 10th, '61, at Richmond.

Hebrew, Peter, e. Sept. 6th, '62; Moore co.

Ingram, Benj., e. Aug. 22d, '61; Anson co.; k. May 31st, '62, at Seven Pines.

Ingram, W. D., e. Aug. 22d, '61; Anson co.; k. Sept. 17th, '63, at Sharpsburg.

Jones, J. W., e. Sept. 6th, '62; Henderson co.

Kelly, L. M., e. Aug. 22d, '61; Anson co.; d. May 3d, '62, at Richmond.

Knight, Jas., e. Aug. 22d, '61; Anson co.; d. May 30th, '63, of w. received at Chancellorsville.

Killiam, A. L., e. Sept. 6th, '62; Catawba co.; d. Feb. 4th, '63, at Lynchburg.

Keiver, Milton, e. Sept. 6th, '62; Catawaba co.

Killiam, J. A., e. Sept. 6th, '62; Lincoln co.; w. at Chancellorsville.

King, W. H., e. Sept. 6th, '62; Henderson co.; pr. at Gettysburg.

Kerby, C. C., e. May 12th, '61; Anson co.; w. and pr. at Gettysburg; p. Sergeant.

Longeryer, Paul, e. Sept. 6th, '62; Catawba co.; d. May 12th, '63, at Lynchburg.

Lee, William, e. May 22d, '61; Anson co.; k. July 1st, '63, at Gettysburg.

Lockhart, James T., e. May 22, '61; Anson co.; w. and pr. at Gettysburg.

Leonard, Caswell, e. March 19th, '64; Anson co.

Mask, M. S., e. Aug. 22d, '61; Anson co.; k. Sept. 17th, '62, at Sharpsburg.

May, John, e. Aug. 22d, '61; Anson co.

Marshall, Jas. C., e. Aug. 22d, '61; Anson co.; p. 1st Lieut. May 10th, '62.

McDuffee, George W., e. Aug. 22d, '61; Anson co.; d. Sept. 10th, '61, at Manassas.

Meadows, Jas. W., e. Aug. 22d, '61; Anson co.; c. Sept. 18th, '64.

Meacham, Coleman, e. Aug. 22d, '61; Anson co.; d. Dec. 31st, '61, near Manassas.

Munlin, Jas. W., e. Aug. 22d, '61; Anson co.; w. near Richmond.

Munnerlyn, John R., e. Aug. 22d, '61; Anson co.; w. at Seven Pines.

Myers, Ransom B , e. Aug. 22d, '61; Anson co.; c. July 1st, '63.

Myers, Stephen H., e. Aug. 22d, '61; Anson co.; dg. Oct. 1st, '62.

Mathis, Daniel, e. Sept. 6th, '62; Catawba co.; dt.

Mull, Jacob, e. Sept. 6th, '62; Catawba co.; w. and pr. at Gettysburg.

Moore, W. M., e. Sept. 6th, '62; Catawba co.; d. Oct. 6th, '62, at Bunker Hill, Va.

Munnerlyn, Robert W., e. Sept. 23d, '62; Montgomery co.; d. Nov. 25th, '62, at Danville.

Mull, D. F., e. Sept. 6th, '62; Catawba co.; d. Jan. 13th, '63, at Fredericksburg.

Munnerlyn, J. W.

Mays, Jackson, e. Sept. 6th, '62; Iredell co.; pr. May 12th, '64.

Parker, J. F., e. Sept. 6th, '62; Catawba co.

Punch, J. L., e. Sept. 6th, '62; Catawba co.; pr. at Gettysburg.

Phillips, Thomas, e. Aug. 22d, '61; Anson co.; w. at Gettysburg; c. Sept. 19th, '64.

Ponds, Tristam T., e. August 22d, '61; Anson co.; d. Aug. 5th, '61, at Manassas.

Proffitt, William R., e. August 22d, '61; Anson co.; dg. October 1st, '62.

Perry, David, e. Sept. 6th, '62; Alexander co.

Redfearn, Alfred, e. August 22d, '61; Anson co.; dg. October 1st, '62.

Redfearn, Berry J., e. August 22d, '61; Anson co.; k. May 29th, '62, in Va.

Redfearn, John, e. August 22d, '61; Anson co.; d. July 6th, '62, of w. at Seven Pines.

Redfearn, James T., e. August 22d, '61; Anson co.; dt. Colonel's Orderly.

Redfearn, Wilson C., e. August 22d, '61; Anson co.; d. October 4th, '61, in Va.

Rutliffe, Dennis, e. August 22d, '61; Anson co.

Richardson, John R., e. August 22d, '61; Anson co.; p. Sergeant; w. at Sharpsburg and Gettysburg.

Ricketts, George W., e. August 22d, '61; Anson co.; dg. Oct. 21st, '63, for disability.

Rushing, Andrew P., e. August 22d, '61; Anson co.; w. at Seven Pines; k. July 1st, '63, at Gettysburg.

Rushing, Horace F., e. August 22d, '61; Anson co.; d. July 10th, '62, in Virginia.

Roach, John J., e. August 22d, '61; Anson co.; dg. June 15th, '62.

Rorie, W. T., e. May 22d, '61; Anson co.

Rushing, Thos., e. May 12th, '61; Anson co.; w. at Chancellorsville; p. Corporal.

Rorie, Thomas, e. May 22d, '61; Anson co.

Ronier, A. J., e. Sept. 6th, '62; Catawba co.; d.

Short, Spencer, e. August 22d, '61; Anson co.; w. at Seven Pines and Chancellorsville.

Sullivan, Isaac, e. August 22d, '61; Anson co.; dg. July 12th, '62, for disability.

Sykes, John T., e. August 22d, '61; Anson co.

Stegall, Henry M., e. May 22d, '61; Anson co.

Saunders, William K., e. May 22d, '61; Anson co.; d. March 17th, '62, in Virginia.

Tarlton, Benjamin, e. May 22d, '61; Anson co.; d. June 31st, '62, of w.

Tarlton, Thomas, e. May 22d, '61; Anson co.; w. at Seven Pines, Sharpsburg and Gettysburg; pr.

Townsend, James S., e. May 22d, '61; Anson co.; pr. at Sharpsburg; w. at Chancellorsville.

Traywick, Samuel, e. May 22d, '61; Union co.; p. Sergeant; d. July 15th, '62, in Virginia.

Tucker, William T., e. May 22d, '61; Anson co.; w. at Chancellorsville; dt. in '64.

Tucker, Lewis, e. May 22d, '61; Anson co.; dg. December 18th, '61, for disability.

Tyson, Jethro, e. May 22d, '61; Anson co.; d. Sept. 15th, '61, at Manassas.

Thornton, J. P., e. Sept. 6th, '62; Henderson co.; d. March 1st, '63, in Va.

Turner, P. L., e. Sept. 6th, '62; Catawba co.; w.

Watkins, W. C., e. May 22d, 61; Anson co.; k. Sept. 17th, '62, at Sharpsburg.

Wall, Samuel C., e. May 22d, '61; Anson co.

Waddell, Wm. R., e. May 22d, '61; Anson co.; dg. Nov. 8th, '61, for disability.

Womble, Thos., e. May 12th, '61; Anson co.

Young, John, e. Sept. 6th, '62; Catawba co.; pr. at Gettysburg.

COMPANY B.

OFFICERS.

George W. Seagle, Captain, cm, May 23d, '61; Lincoln co.

Wesley Hadspeth, Captain, cm. ——; Lincoln co.; p. from ranks; w. at Sharpsburg; k. at Chancellorsville May 3d, '63.

G. W. Hunter, Captain, cm. ——; Lincoln co.; p. from ranks.

Joshua Holbrook, Captain, Lincoln co.; p. from ranks.

T. J. Sagle, 1st Lieut., cm. May 23d, '61; Lincoln co.

M. H. Shuford, 1st Lieut., cm. May 23d, '61; Lincoln co.

Lee Johnson, 2d Lieut., cm. May 23d, '61; Lincoln co.

S. A. Shuford, 2d Lieut., cm. May 23d, '61; Lincoln co.

William R. Sloan, 2d Lieut., cm. May 23d, '61; Mecklenburg co.

M. H. Shuford, 2d Lieut., cm. May 10th, '62; Lincoln co.

W. A. Thompson, 2d Lieut., cm. May 10th, '62; Lincoln co.

M. M. Hines, 2d Lieut., cm. November 20th, '61; Lincoln co.; pr. Sept. 19th, '64.

NON-COMMISSIONED OFFICERS.

A. P. James, 1st Sergeant, e. May 23d, '61; Lincoln co.; dg.

William McCaslin, 2d Sergeant, e. May 23d, '61; Lincoln co.; d. April 23d, '62, at Richmond.

A. J. Seagle, 3d Sergeant, e. May 23d, '61; Lincoln co.; p. Sergeant; w. at Sharpsburg.

John A. Roberts, 4th Sergeant, e. May 23d, '61; Lincoln co.; dg. August, '62.

William R. Sloan, 5th Sergeant, e. May 23d, '61; Mecklenburg co.; p. Lieut.

Levi H. Friger, 1st Corporal, e. May 23d, '61; Lincoln co.; w. at Seven Pines and Gettysburg.

Darrel Friger, 2d Corporal, e. May 23d, '61; Lincoln co.; d. Nov., '61, at Richmond.

M. H. Shuford, 3d Corporal, e. May 23d, '61; Gaston co.; p. 2d Lieut. May 10th, '62.

Martin Ransom, 4th Corporal, e. May 23d, '61; Lincoln co.; p. Sergeant; k. July 1st, '63, at Gettysburg.

PRIVATES.

Black, Alfred, e. March 15th, '62; Gaston co.

Black, Dinkney, e. March 15th, '62; Gaston co.

Baker, Jefferson A., e. March 15th, '62; Gaston co.; k. May 31st, '62, at Seven Pines.

Bysinger, William H., e. March 15th, '62; Lincoln co.; pr. May 3rd, '63, at Chancellorsville.

Bell, Cephas, e. March 15th, '62; Gaston co.; dg. September, '62, for disability.

Blaylock, Alex., e. October 17th, '64; Lincoln co.

Cauble, Peter V., e. March 15th, '62; Lincoln co.

Cauble, Harrison, e. March 15th, '62; Lincoln co.

Cody, Jesse, e. March 15th, '62; Gaston co.; d. July 20th, '62.

Cody, Cameron, e. March 15th, '62; Gaston co.; d. July 20th, '62, of w. received at Seven Pines.

Cross, F. G., e. March 15th, '62; Lincoln co.

Campbell, G. P., e, March 15th, '62; Lincoln co.; d. July 20th, '62.

Campbell, Jere., e. March 15th, '62; Gaston co.; k. May 31st, '62, at Seven Pines.

Carpenter A. S., e. May 23d, '61; Lincoln co.; p. Sergeant; w. at Seven Pines; w. and c. at Gettysburg.

Carpenter, W. D., e. May 23d, '61; Lincoln co.; k. July 1st, '63, at Gettysburg.

Carpenter, David S., e. May 23d, '61; Lincoln co.; k. May 31st, '62, at Seven Pines

Carpenter, Joseph, e. May 23d, '61; Lincoln co.; w. at Seven Pines.

Coon, Henry, e. May 23d, '61; Lincoln co.; k. September 14th, '62, at South Mountain.

Coddell, Henry, e. May 23d, '61; Lincoln co.; d. July 20th, '62.

Campbell, Harman, e. May 23d, '61; Lincoln co.; w. and pr. at Gettysburg; d. July 7th, '62.

Carter, Robert M., e. May 23d, '61; Lincoln co.

Davis, Thomas, e. March 15th, '62; Gaston co.; w. May 3rd, '63, at Chancellorsville.

Edwards, Jasper, e. May 23d, '61; Lincoln co.; d. September, '61, at Charlottsville.

Edwards, Marion, e. May 23d, '61; Lincoln co.

Fentress, Geo. W., e. May 23d, '61; Wake co.; dg. Sept. 4th, '62.

Finger, Wm., e. May 23d, '61; Lincoln co.

Fisher, David, e. May 23d, '61; Lincoln co.; dg. Sept., '62.

Finger, Levi, e. May 23d, '61; Lincoln co.

Gates, William, e. March 15th, '62; Gaston co.

Goodwin, A. M., e. October 25th, '64; Iredell co.

Hobbs, James, e. March 15th, '62; Lincoln co.

Hoke, David A., e. March 15th, '62; Lincoln co.; p. Corporal; w. at Seven Pines; w. July 1st, '63, at Gettysburg.

Henry, Marcus H., e. March 15th, '62; Gaston co.; k. Sept. 17th, '62, at Sharpsburg.

Helms, Hiram, e. March 15th, '62; Lincoln co.; w. June 27th, '62, at Cold Harbor.

Helms, Eli, e. March 15th, '62; Lincoln co.; d. July 15th, '62, of w. received at Cold Harbor.

Hines, M. M., e. March 15th, '62; Gaston co.; p. Lieut. November 20th, '62; c. September 19th, '64.

Haynes, William, e. March 15th, '62; Gaston co.; dg. June 15th, '62, for disability.

Haynes, Marcus, e. March 15th, '62; Lincoln co.; k. May 31st, '62, at Seven Pines.

Halliman, Oliver C., e. March 15th, '62; Gaston co.; dg. January 12th, '62.

Heavener, Nicholas, e. March 15th, '62; Lincoln co.; dg. June 25th, '62, for disability.

Heavener, Frederick, e. March 15th, '62; Lincoln co.

Harrell, Wm. H., e. March 15th, '62; Lincoln co.; d. August 10th, '62, in Va.

Hope, Thomas L., e. March 15th, '62; Lincoln co.

Hoke, Eli, e. August 11th, '62; Lincoln co.; k. May 3d, '63, at Chancellorsville.

Hoke, Marcus, e. January 1st, '63; Lincoln co.; p. Corporal; k. July 1st, '63, at Gettysburg.

Hanser, Levi, e. March 5th, '63; Lincoln co.; k. July 1st, '63, at Gettysburg.

Hoyle, J. F., e. March 5th, '63; Gaston co.; k. May 3d, '63, at Chancellorsville.

Holdbrook, Joshua, e. August 11th, '62; Lincoln co.; p. Captain of Com. B; w. at Seven Pines.

Hunter, G. W., e. April, '—; Lincoln co.; p. Captain; d. July 16th, '63, of w. at Charlotte, N.C.

Harrell, A. T., e. October 17th, '64; Lincoln co.

Harris, Rufus, e. May 23d, '61; Lincoln co.

Hoke, W. M., e. May 23d, '61; Lincoln co.; d. April 20th, '62.

Holdbrooks, John, e. May 23d, '61; Lincoln co.; dg. February, '62, for disability.

Hendrick, Andrew, e. May 23d, '61; Lincoln co.; w. at Chancellorsville.

Hollman, George B., e. May 23d, '61; Lincoln co.; d. May 12th, '63, of w. received at Chancellorsville.

Helms, John, e. May 23d, '61; Lincoln co.; dg. October, '62, for w. received at Seven Pines.

Helms, Jacob, e. May 23d, '61; Lincoln co.; w. May 31st, '62; at Seven Pines.

Haynes, Robert, e. May 23d, '61; Lincoln co.

Hull, John, e. May 23d, '61; Lincoln co.; w. and pr. July, '63, at Gettysburg.

Hull, C. P., e. May 22d, '62; Lincoln co.; dg. June, '62, for disability.

Hull, Major, e. May 23d, '61; Lincoln co.

Harrison, William, e. May 23rd, '61; Tennessee; w. at Seven Pines.

Hobbs, Caleb, e. May 23rd, '61; Lincoln co.; pr. at South Mountain; w. at Chancellorsville.

Hartzoe, Daniel M., e. May 23rd, '61; Lincoln co.

Hayle, Andrew, e. May 23rd, '61; Lincoln co.

Hadspeth, Wesley, e. September 1st, '61; Lincoln co.; p. Captain; w. at Sharpsburg; k. May, '63, at Chancellorsville.

Kistler, W. H., e. May 23rd, '61; Lincoln co.; dg. December, '62, for w. received at Seven Pines.

Kistler, Noah, e. May 23rd, '61; Lincoln co.; k. July 1st, '63, at Gettysburg.

Kirksey, Albert, e. May 23rd, '61; Lincoln co.; dg. December, '62, for w. received at Seven Pines.

Keener, Franklin, e. May 23rd, '61; Lincoln co.; w. and pr. at Sharpsburg and Gettysburg.

Lambeth, A. M., e. May 23rd, '61; Lincoln co.

Leonard, Caleb, e. May 23rd, '61; Lincoln co.

Lutz, Daniel A., e. March 15th, '62; Lincoln co.; w. at Sharpsburg; pr. at Gettysburg.

Le-Fevre, William, e. March 15th, '62; Lincoln co.; w. at Seven Pines and dg.

Loftin, Langford, e. November 1st, '64; Lincoln co.

Mosteller, William P., e. March 15th, '62; Lincoln co.; d. August 19th, '62, in Virginia.

Morrison, Nelson, e. March 15th, '62; Lincoln co.; k. September 14th, '62, at South Mountain.

Martson, Daniel, e. May 25th, '61; Lincoln co.

Moore, A. D., e. May 23rd, '61; Lincoln co.; w. at Seven Pines.

Mace, John, e. May 23rd, '61; Lincoln co.; w. at Malvern Hill.

Neal, Peter, e. March 15th, '62; Lincoln co.; k. May 31st, '62, at Seven Pines.

Oaks, John, e. May 23rd, '61; Lincoln co.; dg. August, '62.

Pelt, Marcus, e. May 23rd, '61; Lincoln co.; d. December, '61, at Richmond.

Peal, Giles, e. October 17th, '64; Lincoln co.

Pelt, Jacob, e. May 23rd, '61; Lincoln co.; w. May 31st, '62, at Seven Pines.

Parker, David, e. May 23rd, '61; Lincoln co.; pr. July 19th, '63.

Potter, John, e. May 23rd, '61; South Carolina.

Quickle, Jacob, e. May 23rd, '61; Lincoln co.; k. May 31st, '62, at Seven Pines.

Rhodes, Daniel F., e. May 23rd, '61; Lincoln co.; w. May 31st, '62, at Seven Pines.

Ramsour, Milton, e. May 23rd, '61; Lincoln co.; w. May 31st, '62, at Seven Pines; p. Corporal.

Ramsour, John, e. September 1st, '61; Lincoln co.; dg. August, '62.

Ramsour, M. S., e. May 23rd, '61; Lincoln co.; d. April 29th, '62, at Yorktown.

Reep, Peter, e. May 23rd, '61; Lincoln co.; k. May 3rd, '63, at Chancellorsville.

Reynolds, Jacob, e. May 23rd, '61; Gaston co.; k. May 31st, '62, at Seven Pines.

Robinson, Joseph R., e. May 23d, '61; Lincoln co.; k. May 3d, '63, at Chancellorsville.

Ransom, P. W., e. May 23d, '64; Lincoln co.; c. at Cedar Creek October 19th, '64.

Rash, Noah, e. September 25th, '63.

Rhodes, David F., e. May 23d, '61; Lincoln co.

Ransom, E. D., e. May 15th, '64; Lincoln co.

Ramsour, Pinkney, e. March 15th, '62; Lincoln co.; w. at Seven Pines and Chancellorsville.

Rudisill, Martin, e. March 15th, '62; Lincoln co.; k. July 1st, '63, at Gettysburg.

Rosh, Melvin, e. March 15th, '62; Lincoln co.; w. Sept. 17th, '62; tr.

Seagle, Charles, e. March 15th, '62; Lincoln co.; d. February 10th, '63, at Castle Thunder.

Seagle, Polk D., e. March 15th, '62; Lincoln co.; p. Corporal.

Seagle, Marcus, e. March 15th, '62; Lincoln co.; dg. Jan. 15th, '62.

Seronce, Abram, e. March 15th, '62; Lincoln co.

Shuford, Jacob M., e. March 15th, '62; Gaston co.; pr. July 18th, '63, at Front Royal.

Sherrill, Leander, e. March 15th, '62; Lincoln co.; dg. January 15th, '62, for disability.

Sherrill, Nelson M., e. March 15th, '62; Lincoln co.

Sumerow, Geo. W., e. March 15th, '62; Gaston co.; d. April 26th, '63, in Va.

Shuford, M. H., e. ——; Lincoln co.; p. 1st Lieut.; c. at Gettysburg July 1st, '63.

Seagle, Marion, e. January 10th, '63; Lincoln co.; k. May 3d, '63, at Chancellorsville.

Scroner, Ephraim, e. March 5th, '63; Lincoln co.; w. May 3d, '63, at Chancellorsville.

Smith, Thomas, e. March 5th, '63; Lincoln co.; pr. July 1st, '63, at Gettysburg.

Seagle, M. V., e. May 23d, '61; Lincoln co.; k. May 3d, '63, at Chancellorsville.

Seagle, John, e. May 23d, '61; Lincoln co.; k. July 1st, '63, at Gettysburg.

Seagle, George, e. May 23d, '61; Lincoln co.; k. July 1st, '63, at Gettysburg.

Seagle, Noah W., e. May 23d, '61; Lincoln co.; d. July 12th, '62, in prison.

Seagle, Philip, e. May 23d, '61; Lincoln co.; w. and pr. at Gettysburg.

Shuford, James A., e. May 23d, '61; Lincoln co.; dg. September 1st, '62.

Shitle, Lawson, e. May 23d, '61; Lincoln co.; d. October 10th, '62, in Virginia.

Shitle, John A., e. May 23d, '61; Lincoln co.; k. May 3d, '63, at Chancellorsville.

Stuart, Franklin L., e. May 23d, '61; Lincoln co.; w. May 3d, '63, at Chancellorsville.

Sims, Harrison, e. May 23d, '61; Lincoln co.; pr. July 1st, '63, at Gettysburg.

Sains, Elam, e. May 23d, '61; Lincoln co.; k. May 31st, '62, at Seven Pines.

Smith, Andrew, e. May 23d, '61; Lincoln co.; w. June 27th, '62, at Cold Harbor.

Thompson, W. A., e. May 23d, '61; Lincoln co.; p. 2d Lieut. May 10th, '62.

Turbyfield, Marcus, e. May 23d, '61; Lincoln co.; p. Sergeant; pr. at Gettysburg.

Whetstine, David, e. May 23d, '61; Lincoln co.

Wilson, Jacob K., e. May 23d, '61; Lincoln co.; w. May 31st, '62, at Seven Pines.

White, Amos, e. May 23d, '61; Lincoln co.; w. at Mechanicsville; k. July 1st, '63, at Gettysburg.

Whitworth, George R., e. May 23d, '61; Gaston co.

Warlick, Henry, e. May 23d, '61; Lincoln co.; k. May 31st, '62, at Seven Pines.

Wilson, Jacob K., e. May 28th, '61; Lincoln co.

Wilson, La-Fayette, e. August 11th, '62; Lincoln co.; d. Jan. 10th, '63, in Virginia.

Whitworth, W. R., e. March 15th, '62; Lincoln co.; w. May 3d, '63, at Chancellorsville.

White, James G., e. March 15th, '62; Lincoln co.; pr. at Gettysburg.

Williams, Marcus, e. March 15th, '62; Lincoln co.

Yonnt, Ambrose, e. May 23d, '61; Lincoln co.; d. October, '61, at Richmond.

Yonnt, Eli, e. May 23d, '61; Lincoln co.; dg. in '61 for disability.

COMPANY C.

OFFICERS.

C. J. Cochrane, Captain, cm. May 27th, '61; Montgomery co.

E. J. Christian, Captain, cm. May 10th, '62; Montgomery co.; p. Major and k. May 31st, '62, at Seven Pines.

A. F. Scarborough, Captain, cm. May 10th, '62; Montgomery co.; k. at Seven Pines May 31st, '62.

E. H. Lyon, Captain, cm. May 31st, '62; Granville co.; tr. from Com. E; pr. Sept. 19th, '64.

E. J. Christian, 1st Lieut., cm. May 27th, '61; Montgomery co.; p. and k.

John R. Nicholson, 1st Lieut., cm. May 10th, '62; Montgomery co.

E. J. Garris, 2d Lieut., cm. May 27th, '61; Montgomery co.

G. W. Montgomery, 2d Lieut., cm. May 27th, '61; Montgomery co.

Jeremiah Coggins, 2d Lieut., cm. May 10th, '62; Montgomery co.; pr. at Gettysburg July 1st, '63.

A. F. Saunders, 2d Lieut., cm. May 10th, '62; Montgomery co.

J. P. Leach, 2d Lieut., cm. April 14th, '63; Montgomery co.

NON-COMMISSIONED OFFICERS.

J. P. Leach, 1st Sergeant, e. May 27th, '61; Montgomery co.; p. 2d Lieut. April 14th, '63.

Jeremiah Coggins, 2d Sergeant, e. May 27th, '61; Montgomery co.; p. 2d Lieut. May 10th, '62.

John R. Nicholson, 3d Sergeant, e. May 27th, '61; Montgomery co.; p. 1st Lieut. May 10th, '62.

E. J. Lilly, 4th Sergeant, e. May 27th, '61; Montgomery co.; w. May 3rd, '63, at Chancellorsville.

George T. Bledsoe, 5th Sergeant, e. May 27th, '61; Montgomery co.; d. Aug. 23d, '63, in N. C., of w. received at Seven Pines.

William Harris, 1st Corporal, e. May 27th, '61; Montgomery co.; w. Sept. 17th, '62, at Sharpsburg.

Samuel H. Marion, 2d Corporal, e. May 27th, '61; Montgomery co.; p. Sergeant and w. at Seven Pines and Gettysburg.

Abram Coggins, 3d Corporal, e. May 27th, '61; Montgomery co.; d. Sept., '62, of w. received at Sharpsburg.

Thomas J. Bright, 4th Corporal, e. May 27th, '61; Montgomery co.; k. July 1st, '62, at Malvern Hill.

O. L. Miller, Musician, e. May 27th, '61; Montgomery co.

M. Shaw, Musician, e. May 27th, '61; Montgomery co.

PRIVATES.

Annsucker, A. A., e. May 27th, '61; Montgomery co.; w. at Seven Pines and Chancellorsville.

Andrews, Edmund, e. May 27th, '61; Montgomery co.; dg. June, '62.

Andrews, Whitson W., e. May 27th, '61; Montgomery co.; dg. Jan. 7th, '62, for disability.

Andrews, William H., e. May 27th, '61; Montgomery co.; d. August, '61, in N. C.

Ary, B. T., e. September 3d, '62; Stanly co.; w. at Gettysburg.

Boon, E. C., e. September 3d, '62; Stanly co.

Brown, L., e. September 3d, '62; Stanly co.

Burris, David, e. September 3d, '62; Stanly co.

Burris, J. L., e. September 3d, '62; Stanly co.; d. June 10th, '63, of w. at Chancellorsville.

Burris, William A., e. September 3d, '62; Stanly co.

Brewer, E. S., e. May 27th, '61; Montgomery co.; d. Aug. 22d, '62, at Liberty, Va.

Brewer, James D., e. May 27th, '61; Montgomery co.; d. in Hospital.

Bowden, Thomas, e. May 27th, '61; Montgomery co.; w. at Chancellorsville and Gettysburg.

Boyd, Ira, e. May 27th, '61; Montgomery co.; d. Aug. 9th, '62, at Richmond.

Boyd, Robert, e. March 4th, '63; Montgomery co.; k. July 1st, '63, at Gettysburg.

Brown, Temple, e. May 27th, '61; Montgomery co.; dg. June, '61, for disability.

Christian, John G., e. May 27th, '61; Montgomery co.; dg. Nov., '62.

Campbell, Alexander, e. May 27th, '61; Montgomery co.; d. June 29th, '62, of w. received at Cold Harbor.

Campbell, James P., e. May 27th, '61; Montgomery co.; d. Aug. 5th, '62, in Va.

Campbell, John H., e. May 27th, '61; Montgomery co.; w. at Gettysburg.

Coggins, John M., e. May 27th, '61; Montgomery co.; d. of w. August 8th, '62, at Seven Pines.

Coggins, Jas., e. May 27th, '61; Montgomery co.; k. Sept. 17th, '62, at Sharpsburg.

Caudle, Thomas J., e. May 27th, '61; Montgomery co ; w. June 27th, '62, at Cold Harbor.

Coggin, Jere., p. 2d Lieut., and pr. at Gettysburg, July 1st, '63

Deaton, H., e. Sept. 3d, '64; Iredell co.

Davis, E. F., e. May 27th, '61; Montgomery co.; p. 2d Sergeant.

Davis, Wilson, e. May 27th, '61; Montgomery co.

Davis, Wm. G., e. June 31st, '62; Montgomery co.; w. at Chancellorsville.

Davis, Wiley P., e. March 4th, '63; Montgomery co.; k. May, '63, at Chancellorsville.

Demas, F. W., e. Sept. 9th, '61; Montgomery co.; k. May 31st, '62, at Seven Pines.

Deaton, Hiram, e. August 31st, '62; Cabarrus co.

Epps, Alex., e. May 27th, '61; Montgomery co.; p. Sergeant; w. at Malvern Hill, and pr. at South Mountain.

Freeman, Branson, e. May 27th, '61; Montgomery co.

Futrel, H. Y., e. May 27th, '61; Montgomery co.; d. August 6th, '61, at Richmond.

Fink, John C., e. August 31st, '62; Cabarrus co.; dg. Feb., '63, for disability.

Freece, M. C., e. August 31st, '62; Cabarrus co.; k. at Chancellorsville.

Furr, Israel, e. Sept. 3d, '62; Stanly co.; d. Jan. 31st, '63, at Hanover Junction.

Green, Anderson, e. May 27th, '61; Montgomery co.; w. at Seven Pines and Gettysburg.

Green, Elijah, e. May 27th, '61; Montgomery co.; w. at Chancellorsville.

Green, Calvin, e. May 27th, '61; Montgomery eo.; p. Corporal; pr. at Gettysburg.

Green, John, e. May 27th, '61; Montgomery eo.; d. of w.

Green, John, e. Sept. 3d, '62; Stanly co.; d.

Green, H. N., e. Sept. 3d, '62; Stanly eo.; m. May 12th, '64.

Green, M. V., e. Sept. 3d, '62; Stanly co.; m. May 12th, '64.

Garrett, Henry, e. May 27th, '61; Montgomery co.; d. of w.

Graves, Jesse, e. May 27th, '61; Montgomery co.; d. June, '62, in N. C.

Hall, John L., e. May 27th, '61; Montgomery co.; w. May 31st, '62, at Seven Pines.

Hall, T. W., e. May 27th, '61; Montgomery eo ; w. and pr. at Gettysburg.

Hall, George W., e. May 27th, '61; Montgomery eo.; pr. at Gettysburg.

Hall, Josiah, e. May 27th, '61; Montgomery eo.; dg. Nov. 13th, '61, for disability.

Hall, E. D., e. May 27th, '61; Montgomery eo.; d. October, '61, at Richmond.

Hall, Nelson T., e. May 27th, '61; Montgomery co.; d. September, '61, in Va.

Harris, Brantly, e. May 27th, '61; Montgomery eo.; k. May, '63, at Chancellorsville.

Harris, Arthur C., e. May 27th, '61; Montgomery eo.; d. Feb. 3d, '62, in Virginia.

Harris, Thomas N., e. May 27th, '61; Montgomery eo.; dt.

Harris, Belia F., e. May 27th, '61; Montgomery co.; pr. at Gettysburg.

Harris, J. Ward, e. May 27th, '61; Montgomery co.; p. Sergeant; w. at Gettysburg.

Harris, S. W., e. May 27th, '61; Montgomery co.; k. Sept. 17th, '62, at Sharpsburg.

Hamilton, W. G., e. May 27th, '61; Montgomery co.; dg. August, '62.

Howell, Edward F., e. May 27th, '61; Montgomery eo.; pr. at South Mountain.

Hearne, Joel, e. May 27th, '61; Montgomery co.; w. and pr. at Gettysburg.

Hilliard, Preston, e. May 27th, '61; Montgomery co.; w. at Chancellorsville.

Haywood, John R, e. May 27th, '61; Montgomery co.; d. June 10th, '62, of w. received at Seven Pines.

Hunt, R. K., e. May 27th, '61; Montgomery co.; d. June 10th, '62, of w. received at Seven Pines.

Hutchins, John, e. May 27th, '61; Montgomery co.; d. July 1st, '62, of w. received at Seven Pines.

Hough, J. A., e. May 27th, '61; Montgomery co.

Harris, Wm., e. Sept. 3d, '62; Iredell co.

Hunsucker, A. H., e. May 27th, '61; Montgomery co.

Heathcock, J. M., e. Sept. 3d, '62; Stanly co.; pr. at Gettysburg.

Honeycutt, A. F., e. Sept. 3d, '62; Stanly co.; m. October 19th, '64.

Honeycutt, C. M., e. Sept. 3d, '62; Stanly co.; m. October 19th, '64.

Honeycutt, E. E., e. Sept. 3d, '62; Stanly co.; d. February 13th, '63.

Jenkins, William R., e. May 27th, '61; Montgomery co ; w. at Sharpsburg.

Jordan, Jacob, e. May 27th, '61; Montgomery eo.

Lewis, J. F., e. May 27th, '61; Montgomery co

Lilly, J. Madison, e. May 27th, '61; Montgomery eo.; p. Corporal.

Lyon, E. H., p. Captain and pr. at Winchester Sept. 19th, '64.

Lents, David T., e. Sept. 2d, '62; Stanly co.; d. Jan. 17th, '63, in Virginia.

Lefter, W. R., e. Sept. 3d, '62; Stanly eo.; k. July 1st, '63, at Gettysburg.

Lippard, W. H., e. August 30th, '62; Cabarrus eo.; d. Nov. 29th, '62, in Virginia.

Lippard, Peter, e. August 30th, '62; Cabarrus co.

Litaker, W. N., e. August 30th, '62; Cabarrus co.; m. September 19th, '64.

Little, Jacob, e. August 30th, '62; Cabarrus co.; dg. Jan. 24th, '63, for disability.

Love, James E., e. Sept. 3d, '62; Stanly co.; w. at Chancellorsville.

Leach, Daniel P., e. Sept. 9th, '61; Montgomery co.; d. December 1st, '61, at Charlottesville.

****

Morrison, A. C., e. May 27th, '61; Montgomery co.; w. at South Mountain; dt.

Morrison, James E., e. May 4th, '63; Montgomery co.

Moore, John M., e. May 27th, '61; Montgomery co.

Moore, Ezekiel, e. May 27th, '61; Montgomery co.; d. Oct. 1st, '62, in Va.

Munn, E. B., e. May 27th, '61; Montgomery co.

Martin, James, F., e. May 27th, '61; Montgomery co.; dg. October 1st, '61, for disability.

Mason, Ralph, e. May 27th, '61; Montgomery co.; m. May 12th, '64.

Miller, O., e. May 27th, '61; Montgomery co.

Mills, Robert J., e. May 27th, '61; Montgomery co.; pr. July 1st, '63, at Gettysburg.

Morgan, J. M., e. Sept. 3d, '63; Stanly co.; dt.

Nall, Willis A., e. May 27th, '61; Moore co.; p. Corporal; w. at Cold Harbor and Stevensburg.

Owen, E. H., e. May 27th, '61; Moore co.

Parker, Lewis, e. May 27th, '61; Montgomery co.; w. at Gettysburg.

Parker, S. S., e. May 27th, '61; Montgomery co.; pr. July 1st, '63, at Gettysburg.

Powell, Thomas F., e. May 27th, '61; Richmond co.; p. Sergeant.

Parnell, John, e. August 30th, '62; Iredell co.; m. July 1st, '63.

Parnell, J. N., e. August 30th, '62; Cabarrus co.; dt.

Poteat, J. N., e. Aug. 30th, '62; Cabarrus co.

Pemberton, Thomas, e. May 27th, '61; Montgomery co.: w. at Seven Pines; d. July 8th, '62, in Va.

Robinson, W. C., e. May 27th, '61; Montgomery co.; d. Sept. '61, in Va.

Robinson, W. H. H., e. May 27th, '61; Montgomery co.; d. February 1st, '62, at Manassas.

Robinson, William S., e. May 27th, '61; Montgomery co.; pr. at South Mountain and Gettysburg.

Robinson, John S., e. May 27th, '61; Montgomery co.; w. and pr. at Gettysburg.

Russell, A., e. May 27th, '61; Montgomery co.; d. September 29th, '62, of w. received at Sharpsburg.

Ragsdale, Timothy, e. May 27th, '61; Montgomery co.; d. June 1st, '62, in Virginia.

Rumple, J. N., e. August 27th, '63; Cabarrus co.

Saunders, A. F., e. May 27th, '61; Montgomery co.; p. 2d Lieut. May 10th, '62.

Saunders, James L., e. May 27th, '61; Montgomery co.; w. at Sharpsburg.

Saunders, R. F., e. May 27th, '61; Montgomery co.; dg. November, '62.

Scarborough, A. F., e. May 27th, '61; Montgomery co.; p. Captain May 10th, '62.

Scarborough, William, e. May 27th, '61; Montgomery co.

Scarborough, Wm., e. May 27th, '61; Montgomery co.

Shaw, Martin, e. May 27th, '63; Montgomery co.

Shaw, Martin, e. May 27th, '61; Montgomery co.; e. June 1st, '64.

Smith, William R., e. Sept. 5th, '62; Stanly co.; d. of w. received at Chancellorsville.

Teah, John C., e. May 27th, '61; Montgomery co.; d. in North Carolina.

Tucker, L. M., e. Sept. 5th, '62; Stanly co.; k May 3d, '63, at Chancellorsville.

Tucker, J. T., e. Sept. 5th, '62; Stanly co.; pr. at Gettysburg.

Webb, J. L., e. Sept. 5th, '52; Stanly co.; d.

Williams, A. R., e. May 27th, '61; Montgomery co.; d. December, '62, in Virginia.

Wallace, L. H., e. May 27th, '61; Montgomery co.; d. January, '62, in Virginia.

Wallace, William P., e. May 27th, '61; Montgomery co.; k. July 1st, '63, at Gettysburg.

Woodle, Henson, e. May 27th, '61; Montgomery co.

Wade, C. C., e. May 27th, '61; Montgomery co.; dg. for disability.

White, Wm., e. May 27th, '61; Montgomery co.; dt.

White, Thomas, e. May 27th, '61; Montgomery co.; d. in '62, in Va.

14   **NORTH CAROLINA - 23rd REGIMENT INFANTRY - Roster and History**

****

COMPANY D.

OFFICERS.

Lewis H. Webb, Captain, cm. May 30th, '61; Richmond co.

A. T. Cole, Captain, cm. May 10th, '62; Richmond co.; w. at Sharpsburg and Chancellorsville; w. and pr. at Gettysburg.

James S. Knight, 1st Lieut., cm. May 30th, '61; Richmond co.; k. at Chancellorsville May 3d, '63.

Risden T. Nichols, 1st Lieut., cm. May 10th, '62; Richmond co.; d. in '62.

J. H. Chappell, 1st Lieut., Richmond co.

John W. Cole, 2d Lieut., cm. May 30th, '61; Richmond co.

B. H. Covington, 2d Lieut., cm. May 30th, '61; Richmond co.

W. C. Wall, 2d Lieut., cm. Oct. 17th, '61; Richmond co.

James H. Chappell, 2d Lieut., cm. Oct. 10th, '62; Richmond co.

E. A. McDonald, 2d Lieut., cm. Oct. 10th, '62; Richmond co.

M. O. Strickland, 2d Lieut., Richmond co.

NON-COMMISSIONED OFFICERS.

Wm. C. Covington, 1st Sergeant, e. May 30th, '61; Richmond co.; w. at Chancellorsville; w. and pr. at Gettysburg; r.

Alexander T. Cole, 2d Sergeant, e. May 30th, '61; Richmond co.; p. Captain; w. at Malvern Hill and Chancellorsville; pr.

James W. Kenedy, 3d Sergeant, e. May 30th, '61; Richmond co.; dg. Oct., '61, for disability.

William C. Everett, 4th Sergeant, e. May 30th, '61; Richmond co.; p. A. Q. M.

Calvin B. McKimmon, 5th Sergeant, e. May 30th, '61; Moore co.; dg. in '61, for disability.

Thomas B. Ledbetter, 1st Corporal, e. May 30th, '61; Richmond co.; w. May 31st, '62, at Seven Pines.

Henry J. Carr, 2d Corporal, e. May 30th, '61; Richmond co.; w. at Chancellorsville and July, '63. at Gettysburg.

James D. Shortridge, 3d Corporal, e. May 30th, '61; Richmond co.; tr. to 23d Regiment in '62.

John C. Vesserx, 4th Corporal, e. May 30th, '61; Richmond co.; pr. May 31st, '62, at Seven Pines.

PRIVATES.

Andrews, Elbert, e. Sept. 6th, '62; Stanly co.; k. May, '63, at Chancellorsville.

Beal, Giles, e. Sept. 6th, '62; Lincoln co.

Brinkle, Nicholas, e. Sept. 6th, '61; Rowan co.

Bryant, Angus R., e. May 30th, '61; Robeson co.

Benoist, James, e. May 30th, '61; Richmond co.; w. at Sharpsburg and Gettysburg.

Benoist, Daniel, e. May 30th, '61; Richmond co.; w. at Gettysburg.

Bounds, Wiley Wilson, e. May 30th, '61; Richmond co.; w.

Chappell, James H., e. May 30th, '61; Richmond co.; p. 2d Lieut. May 10th, '62.

Chappell, Parks, e. May 30th, '61; Richmond co.

Coles, William C., e. May 30th, '61; Richmond co.; dg. July 6th, '62.

Curtis, James H., e. May 30th, '61; Moore co.; d. January, '62.

Covington, Jas. M., e. May 30th, '61; Richmond co.; p. Sergeant; dg. July 25th, '62.

Covington, J. W., e. May 30th, '61; Richmond co.; w. at Seven Pines and Malvern Hill.

Covington, William C., e. September 6th, '62; Richmond co.; w. at Chancellorsville.

Covington, J. G., e. May 30th, '61; Rockingham co.

Calicut, Paschal, e. Sept. 6th, '61; Rowan co.

Clifford, Branch G., e. Sept. 6th, '61; Rowan co.

Connor, Lawson B., e. Jan. 27th, '62; Gaston co.; tr. from 28d Regiment.

Coster, B. F., e. May 30th, '61; Richmond co.; dg. Sept., '61, for disability.

Crouch, C. C., e. May 30th, '61; Richmond co.; w. at Cold Harbor and South Mountain.

Cole, Dudley, e. May 30th, '61; Richmond co.; k. July, '63, at Gettysburg.

Duncan, John, e. May 30th, '61; Richmond co.; k. May 12th, '64, at Spottsylvania Court House.

Donnahoe, Franklin, e. May 30th, '61; South Carolina.

Dawkins, John W., e. August 11th, '61; Richmond co.; w. at Seven Pines and Gettysburg.

Edgerson, John, e. Sept. 6th, '62; Rowan co.

Eller, Joshua, e. Sept. 6th, '62; Rowan co.; w. Gettysburg.

Eller, Moses, e. Sept. 6th, '61; Rowan co.; m.

Eller, Richard C., e. Sept. 6th, '63; Rowan co.; d. Nov., '63, at Winchester, Va.

Endy, Wm. C., e Sept. 6th, '61; Rowan co.; d. Nov. 7th, '62, in Va.

Fisher, James C., e. Jan. 21st, '61; Catawba co.; tr. from Com. F and d. April 2d, '63, in Va.

Freeman, Martin C., e. May 30th, '61; Richmond co.; d. in Hospital.

Garrett, Oliver C., e. May 30th, '61; Richmond co.; p. Corporal; dt.

Garrett, John F., e. May 30th, '61; Richmond co.; d. Sept. 25th, '62, of w. received at Malvern Hill.

Gibson, Wm. N., e. May 30th, '61; Richmond co.; p. Sergeant; w. at Gettysburg.

Gilbert, Wm., e. Sept. 6th, '62; Caldwell co.; d. of w.

Gilbert, Willis, e. Sept. 6th, '62; Caldwell co.; d. June 20th, '63, in Va.

Hart, Brittain, e. Sept. 6th, '61; Lincoln co.; m.

Hart, Walton, e. Sept. 6th, '62; Iredell co.; w.

Hart, M., e. Sept. 6th, '61; Lincoln co.

Helton, Joel, e. September 6th, '62; Rutherford co.; dt.

Hill, Henry, e. September 6th, '62; Rowan co.

Hailey, Thomas, e. July 6th, '62; Richmond co.

Hamer, Alfred, e. May 30th, '61; South Carolina; dg. Oct. 28th, '61, for disability.

Hamer, John H., e. May 30th, '61; South Carolina; d. July 22d, '61, in N. C.

Haily, Hiram H., e. May 30th, '61; Richmond co.; d. March 3d, '63, in N. C.

Haily, William L., e. May 30th, '61; Richmond co.; dg. July 5th, '62.

Hall, Isaac, e. May 30th, '61; Richmond co.; dg. December, '61.

Hart, Edwin S., e. May 30th, '61; Richmond co.; p. Corporal; w. and pr. at Chancellorsville.

Jernigan, Wm., e. May 30th, '61; Richmond co.; w. at Seven Pines.

Jernigan, W. H., e. May 30th, '61; Iredell co.; e. Oct. 8th, '62.

Johnson, Alexander, e. August 6th, '61; Richmond co.; d. October 5th, '61, at Manassas.

Knight, James S., e. Sept. 6th, '62; Richmond co.; k. at Chancellorsvile.

Keener, Daniel, e. September 6th, '62; Lincoln co.

Keener, Marcus, e. September 6th, '62; Lincoln co.

Keener, Martin, e. September 6th, '62; Lincoln co.

Keener, Simon, e. September 6th, '62; Lincoln co.; k. May, '63, at Chancellorsville.

Keller, Solomon L., e. September 6th, '61: Rutherford co.

Kile, Milas A., e. Sept. 6th, '62; Rowan co.

King, Lewis, e. September 6th, '62; Rutherford co.

Kirby, Charles, e. September 6th, '62; Lincoln co.; m.

Leak, R. S., e May 30th, '61; Richmond co.; tr. to Com. F June, '62.

Ledbetter, T. B., e. May 30th, '61; Rockingham co.; dt.

Lutrick, Alfred N., e. September 6th, '62; Rowan co.; d. June 12th, '62, in Va.

Long, James W., e. May 30th, '61; Richmond co.; p. Sergeant and d. in '63.

Morrison, Angus, e. May 27th, '61; Richmond co.; w. July 29th, '64.

Mooreman, Edmund C., e. May 30th, '61; Richmond co.; w. May 31st, '62, at Seven Pines and dt.

Mooreman, Thomas D., e. May 30th, '61; Richmond co.; d. July 19th, '62, of w. received at Malvern Hill.

McDonald, E. A., e. May 30th, '61; Richmond co.; p. 2d Lieut. May 10th, '62.

McKinnon, W. B., e. May 30th, '61; Richmond co.; d. Jan. 13th, '62, in Va.

McLean, Hugh C., e. May 30th, '61; Richmond co.; w. at Chancellorsville, and w. and pr. at Gettysburg.

McKinzie, Kenneth, e. May 30th, '61; Richmond co.; k. May, '63, at Chancellorsville; w. and pr. at Gettysburg.

Mason, Pressley, e. May 30th, '61; Richmond co.; p. Sergeant; w. and pr. at 2nd Manassas; pr. at Gettysburg July 1st, '63.

Martin, Roderick B., e. May 30th, '61; Richmond co.; p. Sergeant and d. in Hospital.

Mallock, Atlas T., e. May 30th, '61; Montgomery co.; w. at Gettysburg.

Morrison, Alex., e. May 30th, '61; Richmond co.; dg.

McKinzie, Alexander, e. May 30th, '61; Moore co.; d. Jan. 10th, '62, in Va.

McLean, Solomon, e. May 30th, '61; Richmond co.; w. Sept. 19th, '64.

McKinnon, Hosea, e. May 30th, '61; Richmond co.; c. and d.

McKinnon, Nicholas B., e. May 30th, '61; Richmond co.; w. and pr. at Gettysburg.

Morrison, Malcolm, e. May 30th, '61; Richmond co.; k. May, '63, at Chancellorsville.

McKethan, John G., e. May 30th, '61; Richmond co.; p. Sergeant and w. at Gettysburg.

McKethan, Wm. A., e. Aug. 11th, '61; Richmond co.; k. May 31st, '62, at Seven Pines.

McCaskill, Daniel, e. August 11th, '61; Richmond co.; dg. Sept. 28th, '62.

McKinnon, Daniel, e. August 11th, '61; Richmond co.; d. Feb., 62, in Va.

Martin, Benjamin, e. Sept. 6th, '62; Richmond co.; d. in '62.

McKinnon, Colin, e. Sept. 6th, '62; Richmond co.; d. in '62.

Misenheimer, M. R., e. Sept. 6th, '63; Rowan co.

Martin, Joseph, e. September 6th, '62; Caldwell co.; dg. March 15th, '63, for disability.

Newberry, Wm. W., e. May 30th, '61; Robeson co.; w. at Gettysburg and dt.

Nichols, R. T., e. May 30th, '61; Richmond co.; p. 1st Lieut. May 10th, '62, and d. in '62.

Nicholson, Malcolm C., e. May 30th, '61; Richmond co.; p. Sergeant; d. May 15th, '62, in Virginia.

Nicholson, Peter H., e. May 30th, '61; Richmond co.; p. Sergeant; d. October 16th, '61, at Richmond.

Phillips, W. J., e. May 30th, '61; Richmond co.; k. July 1st, '62, at Malvern Hill.

Powell, Charles P., e. May 30th, '61; Richmond co.; w. at Malvern Hill and Gettysburg.

Peep, Isham, e. May 30th, '61: Richmond co.

Pennell, Thomas, e. September 6th, '62; Caldwell co.

Pennell, William H., e. September 6th, '62; Caldwell co.

Perkins, James, e. September 6th, '62; Caldwell co.

Paul, James, e. August 6th, '61; Richmond co.: dg. Aug., '62.

Porter, Crawford, e. May 30th, '61; Rockingham co.

Rainwaters, James, e. July 25th, '62; South Carolina; w.

Roberts, John, e. September 6th, '62; South Carolina.

Reep, John, e. Feb. 1st, '62; South Carolina; tr. from 11th Regiment in '63.

Smith, Thomas J., e. May 30th, '61; Richmond co.; pr. May 3d, '63.

Smith, Charles C., e. Nov. 4th, '61; Richmond co.; pr. May 3d, '63.

Smith, Charles C., e. Nov. 4th, '61; Rockingham co.; c. July 1st, '63.

Strickland, Jonathan, e. May 30th, '61; Richmond co.; dg. Dec., '61, for disability.

Strickland, Hardy, e. May 30th, '61; Richmond co.

Strickland, Henry, e. Sept. 6th, '62.

Strickland, Milton O., May 30th, '61.

Strickland, Josph, e. Sept. 6th, '62.

Strickland, Milton O., e. May 30th, '61; Richmond co.; p. 2d Lieut. in '62.

Smith, Zebedee R., e. May 30th, '61; Richmond co.

Stoner, Charles W., e. Sept. 6th, '62; Rowan co.

Seals, Thomas, e. July 6th, '62; South Carolina.

Sessoms, Thomas, e. May 30th, '62; Richmond co.

Shurne, Daniel, e. September 6th, '62; Lincoln co.; dt.

Stirewalt, Frank A., e. September 6th, '62; Rowan co.; w. at Chancellorsville; dt.

Sticklett, Henry, e. September 6th, '62; Rutherford co.

Sticklett, Jos., e. September 6th, '62; Rutherford co.

Sneed, John, e. August 15th, '61; Richmond co.

Sanford, Asbury, e. May 30th, '61; Richmond co.; w. at Gettysburg.

Scott, Michael, e. May 30th, '61; Richmond co.; w. May 31st, '62, at Seven Pines.

Tillett, James W., e. May 30th. '61; Granville co.; dg. Nov. 23d, '61, for disability

Thomas, Robert D., e. May 30th, '61; Richmond co.; d. April 5th, '62, at Petersburg.

Thomas, H. T., e. May 30th, '61; Richmond co.; d. Sept. 22d, '61, in Virginia.

Talbot, George W., e. May 30th, '61; Richmond co.; w. at Gettysburg.

Webb, Stephen W., e. May 30th, '61; Richmond co.; w. at Seven Pines; dt.

Webb, Robert Q., e. May 30th, '61; Richmond co.; w. at Malvern Hill; k. July, '63, at Gettysburg.

Wall, William C., e. May 30th, '61; Richmond co.; p. 2d Lieut. Oct. 17th, '61.

Watkins, Willis, e. May 30th, '61; Richmond co.; d. May 20th, '62, in Virginia.

Wall, Henry C., e. May 30th, '61; Richmond co.; p. Sergeant; dg.

Walker, George, e. September 6th, '62; Caldwell co.; pr.

COMPANY E.

OFFICERS.

T. J. Horner, Captain, cm. June 5th, '61; Granville co.

B. F. Bullock, Captain, cm. ——; Granville co.

E. E. Lyon, 1st Lieut., cm. June 5th, '61; Granville co.; r. August 15th, '61.

T. W. Moore, 1st Lieut., cm. Aug. 15th, '61; Granville co.

J. H. Mitchell, 2d Lieut., cm. June 5th, '61; Granville co.

A. D. Peace, 2d Lieut., cm. June 5th, '61; Granville co.

R. V. Minor, 2d Lieut., cm. Sept. 25th, '62; Granville co.

E. H. Lyon, 2d Lieut., cm. Nov. 12th, '61; Granville co.; tr. as Captain of Company C.

B. F. Bullock, 2d Lieut., cm. Dec. 6th, '61; Granville co.

J. T. Bullock, 2d Lieut., cm. May 10th, '62; Granville co.; pr. May 12th, '64.

A. S. Webb, 2d Lieut., cm. May 10th, '62; Granville co.

NON-COMMISSIONED OFFICERS.

E. H. Lyon, 1st Sergeant, e. June 5th, '61; Granville co.; p. 2d Lieut. Nov. 12th, '61; p. and tr. as Captain of Com. C.

B. F. Bullock, 2d Sergeant, e. June 5th, '61; Granville co.; p. 2d Lieut. December 6th, '61.

E. L. Fleming, 3d Sergeant, e. June 5th, '61; Granville co.; w. at South Mountain and pr. at Gettysburg.; w. July 18th, '64.

J. W. Fleming, 4th Sergeant, e. June 5th, '61; Granville co.; d. May 2d, '62, in Virginia.

W. J. Rogers, 5th Sergeant, e. June 5th, '61; Granville co.; p. and w. at Gettysburg.

J. T. Bullock, 1st Corporal, e. June 5th, '61; Granville co.; p. 2d Lieut. May 10th, '62, and c. May 12th, '64.

E. H. Winston, 2d Corporal, e. June 5th, '61; Granville co.; d. June 18th, '62, of w. received at Seven Pines.

C. W. Bennett, 3d Corporal, e. June 5th, '61; Granville co.; p. Sergeant; w. at Sharpsburg and dt.

R. B. Beasley, 4th Corporal, e. June 5th, '61; Granville co.; dg. Aug. 4th, '62.

PRIVATES.

Atkins, F. C., e. Sept. 5th, '61; Granville co.; w. at South Mountain; dt.

Bowlin, James W., e. June 5th, '61; Granville co.; m. October 19th, '64.

Bowlin, P. P., e. July 8th, '62; Granville co.

Beck, J. J., e. July 8th, '62; Granville co.; d. September 17th, '62, of w. at Sharpsburg.

Beck, R. H., e. July 8th, '62; Granville co.; w. at South Mountain; dt.

Burchett, J. F., e. June 5th, '61; Granville co.

Burchett, Isaac, e. June 5th, '61; Granville co.; dg. September 26th, '62.

Califor, J. H., e. June 5th, '61; Granville co.

Califor, G. W., e. August 29th, '62; Granville co.; tr. from 46th, Regiment; w. at Sharpsburg; k. July 1st, '63, at Gettysburg.

Cash, Elisha, e. June 8th, '61; Granville co.; w. May 31st, '62, at Seven Pines.

Cash, T. J., e. June 5th, '61; Granville co.; tr. to 12th Regiment in '62.

Cash, N. C., e. July 8th, '62; Granville co.

Coley, A. J., e. July 8th, '62; Granville co.

Coley, A. H., e. July 8th, '62; Granville co.; d. July 5th, '63, of w. received at Chancellorsville.

Coley, S. D., e. July 8th, '62; Granville co.

Coley, M. H., e. June 5th, '61; Granville co.; p. Sergeant; k. July 1st, '63, at Gettysburg.

Coley, B. J., e. June 5th, '61; Granville co.; k. Sept. 14th, '62, at South Mountain.

Crews, D. G., e. June 5th, '61; Granville co.; w. at Chancellorsville.

Crews, E. H., e. June 5th, '61; Granville co.; dt.

Chapell, H. R., e. July 7th, '61; Granville co.; pr. September 1st, '64.

Chapell, J. H., e. June 5th, '61; Granville co.; dg. Sept. 9th, '61, for disability.

Clay, J. G., e. July 15th, '61; Granville co.; w. at South Mountain.

Craft, W. W., e. July 8th, '62; Granville co.; dg. Sept. 25th, '62, for disability.

Clark, C. H., e. June 5th, '61; Granville co.

Clark, Samuel, e. July 8th, '62; Granville co.; w. at Chancellorsville.

Dickinson, A. J., e. June 5th, '61; Granville co.

Duke, W. K., e. June 5th, '61; Granville co.

Duke, W. H., e. June 5th, '61; Granville co.; w. at Chancellorsville.

Davis, C. J., e. June 5th, '61; Granville co ; tr. to 30th Regiment.

Duncan, C. J., e. June 8th, '61; Granville co.; d. August 31st, '61, in Virginia.

Ellington, C. B., e. July 6th, '61; Granville co.; w. at Malvern Hill.

Ellington, H. T., e. June 18th, '61; Granville co.; w. at Seven Pines.

Estes, T. M., e. June 5th, '61; Granville co.; p. Sergeant; w. at Gettysburg; c. July 19th, '64.

Evans, S. S., e. July 8th, '62; Granville co.;

Ferrell, J. G., e. June 5th, '61; Granville co.; w. at Mechanicsville and Gettysburg.

Fleming, T. B., e. June 5th, '61; Granville co.

Fleming, J. W., e. August 5th, '62; Granville co.

Fleming, R. H., e. July 5th, '62; Granville co.; c. May 19th, '64.

Fullenwider, H. W., e. June 5th, '61; Granville co.; w. and pr. July 18th, '64.

Goss, J. T., e. June 5th, '61; Granville co.; k. May 3d, '63, at Chancellorsville.

Goss, John, e. July 8th, '62; Granville co.; w. at Gettysburg; m. Sept. 12th, '64.

Gooch, S. H., e. July 8th, '62; Granville co.; dg. November 1st, '62.

Green, N. H., e. June 5th, '61; Granville co.; w. at Gettysburg.

Gooch, H. R., e. July 8th, '62; Granville co.; dt.

Green, J. B., e. June 5th, '61; Granville co.; dg. September 19th, '62.

Grissom, A. M., e. June 5th, '61; Granville co.; k. July 1st, '63, at Gettysburg.

Harp, F. H., e. June 5th, '61; Granville co.

Heathcock, G. W., e. June 5th, '61; Granville co.; k. September 14th, '62, at South Mountain.

Heflin, J. M., e. June 5th, '61; Granville co.; p. Corporal; w. and pr. at Gettysburg.

Hester, W. C., e. June 6th, '61; Granville co.; d. August 31st, '61.

Huddleston, S. D., e. June 5th, '61; Granville co.

Hobgood, D. G., e. June 5th, '61; Granville co.; w. at Seven Pines.

Holder, W. M., e. June 5th, '61; Granville co.; w. at Gettysburg.

Holleman, Henry, e. June 5th, '61; Granville co.; dg. July 20th, '62.

Horner, H. F., e. June 5th, '61; Granville co.; dg July 21st, '61, for disability.

Hudgins, J. C., e. June 5th, '61; Granville co ; dt.

Inscore, John, e. Sept. 6th, '62; Granville co.

Jackson, Jarrett W., e. June 5th, '61; Granville co.; w. at Seven Pines; k. Sept. 14th, '62, at South Mountain.

Jones, W. H., e. June 5th, '61; Granville co.; w.; dg. Dec. 19th, '62.

Joyner, Robert, e. June 5th, '61; Granville co.; k. Sept. 14th, '62, at South Mountain.

Jones, R. A., e. July 20th, '62; Granville co.

Kimball, L. A., e. June 5th, '62; Granville co.; d. June 30th, '62.

Kinton, J. R., e. June 6th, '61; Granville co.; p. Sergeant; k. May 3d, '63, at Chancellorsville.

Keith, W. J., e. July 8th, '62; Granville co.; d. at Staunton, Va.

Lyon, N. C., e. June 5th, '61; Granville co.; w. at Williamsburg; dg. Sept., '63.

Lyon, Z. E., e. July 8th, '62; Granville co.; w. at Gettysburg.

Luttro, H. R., e. Sept. 1st, '61; Granville co.; d. June 26th, '62, in Va.

Laws, Anderson, e. July 8th, '62; Granville co.; w. at South Mountain; pr.

Lawrence, J. H., e. July 8th, '62; Granville co.

Langum, Darrell, e. June 5th, '61; Granville co.

McKinzie, J. D., e. June 5th, '61; Granville co.

Mitchell, W. H., e. June 5th, '61; Granville co.; d. Jan. 12th, '62, in Virginia.

Moore, W. A., e. June 5th, '61; Granville co.; d. October 17th, '61, in Virginia.

Morris, A. H., e. June 5th, '61; Granville co.; tr. to 30th Regiment.

Moss, R. T., e. June 5th, '61; Granville co.; d. September 14th, '61, in Virginia.

Metze, John, e. July 8th, '62; Granville co.; c. May 12th, '64.

Matholomew, John, e. July 8th, '62; Granville co.; d. of w. July 25th, '63, at Gettysburg.

Mason, J. T., e. July 8th, '62; Granville co.; w. and m. at Gettysburg.

Moore, Thomas, e. August 4th, '62; Virginia; pr. Sept. 14th, '62, at South Mountain.

McLery, Thomas, e. July 8th, '62; Granville co.; d. Dec. 17th, '62, at Richmond.

Nance, C. E., e. June 5th, '61; Granville co.; d. September 19th, '61, in Virginia.

Nance, S. W., e. June 5th, '61; Granville co.; k. May 3d, '63, at Chancellorsville.

O'Brian, S. R., e. June 5th, '61; Granville co.; p. Corporal; w. at Hagerstown.

O'Brian, S. J., e. June 5th, '61; Granville co.; k. Sept. 14th, '62, at South Mountain.

O'Brian, T. A., e. June 5th, '61; Granville co.; d. Nov. 19th, '62, at Winchester.

O'Brian, John, e. July 8th, '62; Granville co.; c. July 19th, '64.

Oakley, J. T., e. July 8th, '62; Granville co.; d. Sept. 19th, '62, at Richmond.

Oakley, Ellison, e. July 8th, '62; Granville co.; d. Sept. 3d, '62, in Virginia.

Patterson, J. W., e. June 5th, '61; Granville co.; dg. October 1st, '61.

Peace, J. A., e. June 5th, '61; Granville co.; p. Corporal; m. at Seven Pines.

Perry, J. V. S., e. June 5th, '61; Granville co.; d. Sept. 14th, '61, in Virginia.

Perry, T. H., e. July 8th, '62; Granville co.

Perry, G. W., e. July 8th, '62; Granville co.

Perry, W. W., e. July 8th, '62; Granville co.; c. May 19th, '64.

Peace, G. T., e. July 8th, '62; Granville co.

Paschall, R. M., e. July 8th, '62; Granville co.

Peace, Isaac, e. July 8th, '62; Granville co.; d. Oct. 10th, '62, at Richmond.

Robinson, Major, e. June 17th, '61; Granville co.; d. June 15th, '62, in Virginia.

Rowland, J. H., e. June 5th, '61; Granville co.; p. Corporal; w. at Gettysburg.

Russell, H. C., e. June 5th, '61; Granville co.; pr. May 12th, '64.

Russell, W. H., e. June 1st, '62; Granville co.

Russell, John, e. July 8th, '62; Granville co.; d. August 26th, '62, in Virginia.

Smith, J. H., e. June 5th, '61; Granville co.

Stone, Daniel, e. June 18th, '62; Granville co.

Sandford, Joseph, e. July 8th, '62; Granville co.; k. September 14th, '62, at South Mountain.

Sandford, W. T., e. July 8th, '62; Granville co.; w. at Gettysburg.

Sharron, J. W., e. July 8th, '62; Granville co.; dt.

Sait, J. A., e. November 1st, '62; Granville co.; w. at Chancellorsville.

Vaughan, David, e. June 5th, '61; Granville co.; w. at Seven Pines; dt.

Vaughan, T. Y., e. July 8th, '62; Granville co.; k. July 1st, '63, at Gettysburg.

Vaughan, Dow, e. July 8th, '62; Granville co.

Veasey, Isaac, e. June 5th, '61; Granville co.; d. in '62 in Virginia.

Veasey, Elijah, e. September 1st, '61; Granville co.; w. at Seven Pines.

Veasey, John, e. September 1st, '61; Granville co.; pr. July 1st, '63.

Waller, N. A., e. June 5th, '61; Granville co.; d. June 28th, '62.

Waller, W. J., e. June 5th, '61; Granville co.; w. at South Mountain; m.

Wheeler, B. F., e. June 5th, '61; Granville co.; d. in North Carolina.

Wilkins, W. J., e. July 1st, '61; Granville co.; w. and pr. at Gettysburg.

Williams, S. P., e. Aug. 10th, '62; Granville co.; d. July 5th, '63, of w. received at Gettysburg.

Williams, E., e. July 8th, '62; Granville co.; k. September 14th, '62, at South Mountain.

Wilkinson, Paul, e. July 8th, '62; Granville co.; d. July 9th, '63, in N. C.

Winston, J. A., e. May 2d, '62; Granville co.; k. May 3d, '63, at Chancellorsville.

Webb, A. S., e. Sept. 1st, '62; Granville co.; p. 2d Lieut. May 10th, '62.

———

COMPANY F.

OFFICERS.

M. L. McCorkle, Captain, cm. June 6th, '61; Catawba co.

W. C. Wall, Captain, cm. May 10th, '64; Richmond co.

Jacob H. Miller, 1st Lieut., cm. June 6th, '61; Catawba co.

T. W. Wilson, 1st Lieut., Catawba co.

M. L. Helton, 2d Lieut., cm. June 6th, '61; Catawba co.

R. A. Cobb, 2d Lieut.. cm. June 6th, '61; Catawba co.

G. P. Clay, 2d Lieut., cm. May 10th, '62; Catawba co.

T. W. Wilson, 2d Lieut., cm. May 10th, '62; Catawba co.

W. C. Wall, 2d Lieut.; cm. May 10th, '62; Richmond co.

NON-COMMISSIONED OFFICERS.

L. W. Wilkie, 1st Sergeant, e. June 6th, '61; Catawba co.

H. H. Thornton, 2d Sergeant, e. June 6th, '61; Catawba co.

J. M. Leonard, 3d Sergeant, e. June 6th, '61; tr. to 57th Regiment.

John M. Pruner, 4th Sergeant, e. June 6th, '61

Peter A. Link, 1st Corporal, e. June 6th, '61; Catawba co.; k. in Va.

D. N. McCorkle, 2d Corporal, e. June 6th, '61; Catawba co.; d. July 9th, '62, at Richmond.

Eli F. Rink, 3d Corporal, e. June 6th, '61; Catawba co.; w. at Seven Pines.

Sidney H. Rowe, e. June 6th, '61; Catawba co.; tr. to 12th Regiment.

PRIVATES.

Abernathy, John F., e. June 6th, '61; Catawba co.; w. at Malvern Hill.

Angel, Marcus L., e. March 1st, '62; Catawba co.

Abernathy, S. O., e. Dec. 21st, '64; Stokes co.

Baker, Barton, e. June 6th, '64.

Berry, James M., e. June 6th, '61; Catawba co.; d. December 1st, '62, of w. received at Seven Pines.

Bolch, William H., e. June 6th, '61; Catawba co.

Benfield, Marcus, e. June 6th, '61; Catawba co.

Bynum, James M., e. June 6th, '61; Catawba co.; dg. in '61, for disability.

Beatty, Tyler, e. June 6th, '61; Catawba co.; pr. at Gettysburg.

Bost, W. R. D., e. June 6th, '61; Catawba co.; d. July 9th, '62, of w. received at Seven Pines.

Bumgarner, Miles, e. June 6th, '61; Catawba co.; w. at Fredericksburg.

Bumgarner, H. P., e. June 6th, '61; Catawba co.

Bruce, F. H., e. June 6th, '61; Catawba co.; d. March 31st, '61, at Orange C. H.

Bolch, Israel, e. June 6th, '61; Catawba co.; dg. August 1st, '62, for disability.

Bolch, Anthony, e. June 6th, '61; Catawba co.; w. at Chancellorsville.

Baker, Alfred, e. September 1st, '61; Catawba co.; dg. July 4th, '62, for w. received at Seven Pines.

Burnes, Eli, e. March 10th, '63; Catawba co.; pr. at Winchester.

Cline, William S., e. June 6th, '61; Catawba co.; w. at Chancellorsville.

Cline, Eli, e. June 6th, '61; Catawba co.; k. May 31st, '62, at Seven Pines.

Crawford, W. J., e. June 6th, '61; Catawba co.

Cline, Calvin, e. June 6th, '61; Catawba co.; k. July, '63, at Gettysburg.

Christopher, E. A., e. June 6th, '61; Catawba co.; w. at at Seven Pines; k. on railroad in North Carolina.

Clay, G. P., e. September 1st, '61; Catawba co.; p. 2nd Lieut. May 10th, '62.

Clay, David E., e. March 1st, '62; Catawba co.; w. and d. July 28th, '62, in Va.

Cummings, G. W., e. July 8th, '62; Catawba co.

Deal, J. A., e. June 6th, '61; Catawba co.

Dellinger, e. June 6th, '61; Catawba co.; w. at Chancellorsville.

Dellinger, W. P., e. June 6th, '61; Granville co.

Deitz, J. S., e. June 6th, '61; Granville co.; w. at Gettysburg.

Dagerheart, Pinkney, e. Sept. 1st, '61; Catawba co.

Ekard, Wesley D., e. June 6th, '61; Catawba co.; p. Sergeant; w. at Seven Pines.

Elroy, George, Maryland.

Fisher, Jas. C., e. June 6th, '61; Catawba co; tr. to Com. D.

Fry, John C., e. June 6th, '61; Catawba co.; w. at Seven Pines; k. May, '63, at Chancellorsville.

Fisher, Joel H., e. Sept. 1st, '61; Catawba co.

Gibson, Jas. W., e. June 6th, '61; Catawba co.; dt.

Gross, Daniel, e. July 8th, '62; Catawba co.; c. at Gettysburg.

Heffner, Timothy, e. June 6th, '61; Catawba co.; w. at Gettysburg.

Hayes, Wm., e. June 6th, '61; Catawba co.; d. Sept. 28th, '61.

Holler, D. S., e. June 6th, '61; Catawba co.; d. July 9th, '62.

Hoyle, Wm. C., e. June 6th, '61; Catawba co.; d. Jan. 1st, '62.

Hoover, Jefferson, e. June 6th, '61; Catawba co.

Hartroe, Paul, e. June 6th, '61; Catawba co.; d. Aug. 15th, '61, in Va.

Holler, Gilbert, e. June 6th, '61; Catawba co.

Hoyle, Philip A., e. Oct. 2d, '63.

Hudson, W. H., e. Feb. 16th, '64.

Huffman, M. A., e. June 6th, '61; Catawba co.; w. at Chancellorsville; c. at Winchester.

Hall, John C., e. June 6th, '61; Catawba co.; dg. Sept., '62, for disability.

Holler, M. A., e. June 6th, '61; Catawba co.; w. at Seven Pines; dg. March 13th, '63.

Huffman, L. C., e. Sept. 1st, '61; Catawba co.; d. July 17th, '63.

Helton, M. A., e. March 1st, '62; Catawba co.

****

Helton, A. F., e. Feb. 28th, '63; Catawba co.

Icenhour, M. J., e. June 6th, '61; Catawba co.; w. at Gettysburg and d. Oct. 19th, '64.

Jones, Isaac E., e. June 6th, '61; Catawba co.; c. at Winchester.

Johnson, George, e. March 1st, '62; Catawba co.; w. at Gettysburg.

Jarrett, George, e. March 1st, '62; Catawba co.; w. at Gettysburg and pr. at Winchester.

Johnson, Maxwell, e. March 10th, '63; Catawba co.; d. Nov. 15th, '64.

Killiam, Wm. F., e. June 6th, '61; Catawba co.; w. at Seven Pines and Gettysburg.

Killiam, Wm. L., e. June 6th, '61; Catawba co.; p. 1st Sergeant and c.

Leak, Robert, Richmond co.; tr. to Com. D; dg.

Leonard, D. P., e. June 6th, '61; Catawba co.; d. Oct. 9th, '62.

Lutz, J. S., e. June 6th, '61; Catawba co.; w. at Seven Pines and near Richmond.

Lofton, Eli, e. June 6th, '61; Catawba co.; w. at Gettysburg.

Lofton, Pinkney, e. June 6th, '61; Catawba co.; d. Sept. 15th, '61, in Va.

Lofton, William, e. June 6th, '61; Catawba co.; d. Oct. 20th, '61.

Lael, Alexander, e. June 6th, '61; Catawba co.; dg.

Lael, Lawson, Oct. 13th, '63.

Moore, George A., e. June 6th, '61; Catawba co.

Michael, Noah, e. June 6th, '61; Catawba co.; d. July 16th, '62.

Martin, M. P., e. June 6th, '61; Catawba co.; k. July, '63, at Gettysburg.

Mitchell, Thomas, e. June 6th, '61; Catawba co.; d. Sept. 26th, '61, in Va.

Mays, William, e. June 6th, '61; Catawba co.

Masteller, Lawson, e. June 6th, '61; Catawba co.

McGinness, Albert, e. June 6th, '61; Catawba co.; d. June 1st, '62, of w. received at Seven Pines.

McNeill, George C., e. June 6th, '61; Catawba co.

Miller, Robert, e. June 6th, '61; Catawba co.; dg. Oct. 18th, '62.

Miller, John R., e. June 6th, '61; Catawba co.

Miller, J. M., e. April 3d, '64; Catawba co.; c. May 13th, '64.

Mosteller, J. B., e. March 1st, '61; Catawba co.; d. May 16th, '62.

McCorkle, F. M., e. June 6th, '61; Catawba co.; d. June 17th, '62, in Va.

Marshall, E. W., e. July 8th, '62; Catawba co.; d. Feb. 2d, '63.

Miller, Wesley, e. July 4th, '62; Catawba co.

Pool, James L., e. June 6th, '61; Catawba co.

Pool, John, Catawba co.; tr. from 12th Regiment; d. Aug. 16th, '62.

Parker, Jacob, e. June 6th, '61; Catawba co.; m.

Propst, A. G., e. Sept. 1st, '61; Catawba co.; p. Sergeant; w. at Chancellorsville.

Parker, Albert, e. March 1st, '62; Catawba co.

Propst, Jno. H., e. March 21st, '62; Catawba co.; dg. Oct. 20th, '63, for disability.

Payne, J. S., e. July 8th, '62; Surry co.; c. July 10th, '64.

Pool, Alexander, e. Jan. 16th, '63; Catawba co.; k. May, '63, at Chancellorsville.

Reinhardt, E. F., e. June 6th, '61; Catawba co.; k. July, '63, at Gettysburg.

Rink, Geo. F., e. June 6th, '61; Catawba co.; w. at Gettysburg.

Ramsey, Daniel, e. March 1st, '62; Catawba co.; w. at Gettysburg and near Richmond.

Reinhardt, Abraham, e. July 8th, '62; Catawba co.; k. July, '63, at Gettysburg.

Riggs, Wiley, e. July 8th, '62; Surry co.; w. at Gettysburg; pr.

Reinhardt, Levi, e. March 10th, '63; Catawba co.

Reinhardt, Elias, e. March 10th, '63; Catawba co.; d. June 2d, '63, of w. received at Chancellorsville.

Spencer, Daniel, e. June 6th, '61; Catawba co.; d. March 15th, '62, in N. C.

Seitz, Julius, e. June 6th, '61; Catawba co.

Shell, William D., e. June 6th, '61; Catawba co.; w. at Chancellorsville and Gettysburg.

Seitzer, John F., e. June 6th, '61; Catawba co.

Shell, J. H., e. June 6th, '61; Catawba co.; w. at Seven Pines.

Seitz, David N., e. June 6th, '61; Catawba co.; d. July 7th, '62, of w. received near Richmond.

Sigman, C. C., e. June 6th, '61; Catawba co.

Shuford, Philip, e. March 1st, '62; Catawba co.

Shuford, Solomon, e. March 1st, '62; Catawba co.

Simpson, John, e. July 8th, '62; Cabarrus co.

Suther, David S., e. July 8th, '62; Cabarrus co.

Suther, John A., e. July 8th, '62; Cabarrus co.; d. May 11th, '63, of w. received at Chancellorsville.

Seitz, G. L., e. March 1st, '62; Catawba co.; d. August 3rd, '63, of w. received at

Speagle, Philip, e. Oct. 13th, '63; c. Oct. 19th, '64.
     Gettysburg.

Scornce, Wm. A., e. June 6th, '61; Catawba co.; w. at Seven Pines: pr.

Sigman, M. E., e. June 6th, '61; Catawba co.; d. July 1st, '62, in Va.

Sigman, G. P., e. June 6th, '61; Catawba co,; d. Dec. 4th, '61, in Va.

Shuford, Able A., e. June 6th, '61; Catawba co.; p. Sergeant; w. below Richmond.

Smith, W. H., e. June 6th, '61; Catawba co.; d. Nov. 20th, '61.

Seagle, Adam, e. June 6th, '61; Catawba co.; w. at Gettysburg.

Towell, W. A., e. June 6th, '61; Catawba co.; k. May, '63, at Chancellorsville.

Warlick, G. W., e. June 6th, '61; Catawba co.; w. at Chancellorsville.

Warlick, W. T., e. June 6th, '61; Catawba co.; p. Sergeant; w. at Gettysburg and near Richmond.

Whitener, D. L., e. June 6th, '61; Catawba co. k. below Richmond.

Wingate, Albert, e. June 6th, '61; Catawba co.; d. July 13th, '62.

Wilson, T. W., e. June 6th, '61; Catawba co.; p. 2nd Lieut. May 10th, '62.

Whitener, G. W., e. September 1st, '61; Catawba co.; d. September 20th, '62, at Shepherdstown.

Weaver, John, e. March 1st, '62; Catawba co.

Weaver, J. S., e. July 8th, '62; Catawba co.

Whistenhunt, William, e. March 1st, '61; Catawba co.; pr. May 12th, '64.

Walker, James S., e. July 8th, '62; Catawba co.

Wilcoxen, J. B., e. July 8th, '62; Catawba co.

Wall, W. C., e. May 25th, '63; Richmond co.; p. Captain May 10th, '64; tr. as Lieut. from Com. D.

Warlick, M. H., e. Feb. 28th, '63; Catawba co.; w. at Gettysburg.

Workman, David, e. March 5th, '62; Catawba co.; w. at Gettysburg.

Workman, Daniel, e. March 5th, '63; Martin co.; dg.

Whitener, Newton, e. March 10th, '62; Catawba co.; w. at Chancellorsville and pr.

Yoder, A. M., e. Sept. 1st, '61; Catawba co.; w. thrice.

Yoder, Robert J., e. Sept. 16th, '63.

COMPANY G.

OFFICERS.

C. C. Blacknall, Captain, cm. June 11th, '61; Granville co.; p Major May 31st, '62, and Colonel August, '63; mortally w. September 19th, '64.

I. J. Young, Captain, cm. May 31st, '62; w. May 31st, '62, at Seven Pines; r. August, '62.

T. J. Crocker, Captain, cm. August 15th, '62; Granville co.; w., disabled and r.

James A. Breedlove, Captain, cm. in '64; Granville co.; w.; p. from 1st Lieut.

Isaac J. Young, 1st Lieut., cm. June 11th, '61; Granville co.; p., w. and r.

T. J. Crocker, 1st Lieut., cm. May 31st, '62; Granville co.; p., w. and r.

J. A. Breedlove, 1st Lieut., cm. August 15th, '62; Granville co.; p. and w.

Washington F. Overton, 1st Lieut., cm. in '64; Granville co.

G. W. Kittrell, 2nd Lieut., cm. June 11th, '61; Granville co.

Vines E. Turner, 2nd Lieut., cm. June 11th, '61; Granville co.; p. Adjutant May 10th, '62; w. at Gaines' Mill June 27th, '62; p. A. Q. M. in '63.

T. J. Crocker, 2nd Lieut., cm., May 10th, '62; Granville county; p.

William P. Gill, 2nd Lieut., cm. May 10th, '62; Granville co.; p. from Sergeant Major in '63; k. at Gettysburg.

W. F. Overton, 2nd Lieut., cm. August 15th, '62; Granville co.; p. and k.

J. A. Breedlove, 2nd Lieut., cm. August 15th, '62; Granville co.; p. and w.

C. W. Champion, 2nd Lieut., cm. Nov. 1st, '62; Granville co.

NON-COMMISSIONED OFFICERS.

T. J. Crocker, 1st Sergeant, e. June 11th, '61; Granville co.; p. 2d Lieutenant May 10th, '62.

J. A. Breedlove, 2d Sergeant, e. June 11th, '61; Granville co.; p. 2d Lieutenant August 15th, '62.

W. J. Hinton, 3d Sergeant, e. June 11th, '61; Granville co.; w. July, '63, at Gettysburg.

W. F. Overton, 4th Sergeant, e. June 11th, '61; Granville co.; p. 2d Lieutenant August 15th, '62.

S. S. Hicks, 5th Sergeant, e. June 11th, '61; Granville co.; w. at Mechanicsville; d. November 21st, '62.

Robert T. Champion, 1st Corporal, e. June 11th, '61; Granville co.; w. at Chancellorsville.

James P. Hunt, 2d Corporal, e. June 11th, '61; Granville co.; dg. for disability.

K. W. Coghill, 3d Sergeant, e. June 11th, '61; Granville co.; w. at Sharpsburg.

W. A. Belvin, 4th Corporal, e. June 11th, '61; Granville co.; dg. for disability.

PRIVATES.

Adams, Nathaniel, e. June 11th, '61; Granville co.; w. at Seven Pines; m. May 12th, '64.

Allen, C. E., e. July 8th, '62; Granville co.; dg. September 8th, '62, for disability.

Ball, George N., e. June 11th, '61; Granville co.; m. in action May 9th, '64.

Ball, Edwin, e. June 11th, '61; Granville co.

Bailey, Alexander, e. June 11th, '61; Granville co.; w. at Gettysburg.

Barker, R. B., e. March 15th, '62; Granville co.; d. April, '62, in Virginia.

Best, F. K., e. June 11th, '61; Granville co.; c. Sept. 19th, '64.

Breedlove Bennett, e. March 15th, '62; Granville co.

Breedlove, Randal, e. March 15th, '62; Granville co.; d. Nov., '62.

Breedlove, J. H., e. July 8th, '62; Granville co.; dt.

Breedlove, David, e. June 11th, '61; Granville co.; dg.

Breedlove, J. A.

Brinkley, S. W., e. March 15th, '62; Granville co.; d. June 5th, '62.

Ball, Thomas, e. June 11th, '61; Granville co.; dg. for disability.

Bobbitt, C. J., e. July 8th, '62; Granville co.; m. in action July 1st, '63.

Burrows, J. A., e. June 11th, '61; Granville co.; dt.

Champion, C. W., e. June 11th, '61; Granville co.; p. 2d Lieut. Nov., '62.

Champion, Alexander, e. June 11th, '61; Granville co.; dg. for disability.

Cheatham, W. H., e. March 15th, '62; Granville co.

Coghill, J. F., e. March 15th, '62; Granville co.

Coghill, Joseph, e. June 11th, '61; Granville co.; d. Sept., '62, in Va.

Coghill, J. N., e. June 11th, '61; Granville co.; d. Sept., '62, in Va.

Coghill, K. W., e. July 11th, '61; Granville co.; dt.

Centon, W. J., e. June 11th, '61; Granville co.

Cheatham, T. G., e. July 8th, '62; Granville co.; m. in action July 1st, '63.

Couray, Thomas, e. August 12th, '62; Virginia.

Clark, H. F., e. June 11th, '61; Granville co.

Daniel, Rufus, e. June 11th, '61; Granville co.; w. at Chancellorsville.

Davis, George R., e. June 11th, '61; Granville co.; m. in action July 20th, '64.

Debnam, J. B., e. July 3d, '62; Granville co.; dg. August 12th, '62.

Davis, W. O., e. March 15th, '62; Granville co.

Davis, V. J., e. June 11th, '61; Granville co.; dg. for disability.

Dement, W., e. June 11th, '61; Granville co.; w. at Seven Pines.

Dement, George A., e. June 11th, '61; Granville co.; w. at Seven Pines and Gettysburg.

Dement, Lewis, e. June 11th, '61; Granville co.; d. September, '61.

Dickinson, A. W., e. June 11th, '61; Granville co.; w. at Cold Harbor; d. April
26th, '63.

Dickinson, W. T., e. June 11th, '61; Granville co.; w. at Seven Pines; dg. for disability.

Dickinson, S. T., e. March 16th, '62; Granville co.; m. in action May 19th, '64.

Dickinson, Edwin, e. March 15th, '62; Granville co.

Dunn, Thomas A., e. June 11th, '61; Granville co.; dg. December 11th, '62.

Dunn, B. A., e. March 15th, '62; Granville co.; w. at Malvern Hill; d. June 3d,
'63, of w. received at Chancellorsville.

Egerton, James T., e. June 11th, '61; Granville co.; w. at Williamsburg.

Falkner, William, e. June 11th, '61; Granville co.; m. in action May 1st, '64.

Falkner, Kinley, e. June 11th, '61; Granville co.; d. Sept., '61, in Virginia.

Falkner, George H., e. June 11th, '61; Granville co.; m. in action July 1st, '63.

Falkner, Allen, e. June 11th, '61; Granville co.

Finch, J. D., e. June 11th, '61; Granville co.

Finch, Jordan, e. March 15th, '62; Granville co.; m. in action July 1st, '63.

Finch, Solomon, e. July 8th, '62; Granville co.

Finch, W. S., e. March 15th, '62; Granville co.

Fleming, M. R., e. June 11th, '61; Warren co.; dg. Oct. 4th, '62, for disability.

Floyd, J. L., e. June 11th, '61; Granville co.; k. May 31st, '62, at Seven Pines.

Floyd, Samuel K., e. March 15th, '62; Granville co.; d. in N. C.

Floyd, J. W. S., e. March 15th, '62; Granville co.; d. Nov. 14th, '61.

Fuller, F. M., e. June 11th, '61; Granville co.; dg. March 7th, '63.

Fuller, Arthur, e. March 15th, '62; Granville co.; dg. July 31st, '63.

Fuller, A. E., e. July 8th, '62; Granville co.; dt.

Fuller, E. A., e. June 11th, '61; Granville co.

Grissom, Charles, e. March 15th, '62; Granville co.; d. in Hospital.

Gill, James A., e. June 11th, '61; Franklin co.; w. at Williamsburg and dg.

Gill, W. P., e. June 11th, '61; Franklin co.; p. 2d Lieut. May 10th, '62.

Glasgow, Allen, e. March 15th, '62; Granville co.; w. at Seven Pines.

Gooch, Emmet, e. March 15th, '62; Granville co.; dg. July 24th, '62.

Grissom, Hilliard, e. March 15th, '62; Granville co.; d. in Hospital.

Grissom, G. W., e. March 15th, '62; Granville co.; d. in Hospital.

Harris, Willis, e. June 11th, '61; Warren co.; m. in action July 1st, '63.

Harris, John, e. June 11th, '61; Warren co.; w. and pr. at Gettysburg.

Hedgepeth, J. R., e. March 15th, '62; Granville co.; d.

Hedgepeth, William, e. December 16th, '63; Wake co.

Harp, W. H., e. June 11th, '61; Granville co.; m. in action May 9th, '64.

Harp, John, e. March 15th, '62; Granville co.

Harris, A. K. P., e. March 15th, '62; Warren co.; w. and m. at Gettysburg.

Harp, Fulton, e. March 15th, '62; Granville co.; d.

Hicks, Samuel A., e. June 11th, '61; Granville co.; w. at Sharpsburg; d.

Hicks, B. W., e. July 8th, '62; Granville co.; dg. Dec. 6th, '62.

Holmes, M. H., e. June 11th, '61; Granville co.; w. at Seven Pines and Gettysburg.

Hicks, J. W., e. July 8th, '62; Granville co.

Hicks, John S., e. March 15th, '62; Granville co.

Hays, S. M. S., e. July 8th, '62; Granville co.

Hudson, C. W., e. June 11th, '61; Granville co.; dg.

Hudson, James, e. June 11th, '61; Granville co.; dg. for disability.

Hunt, E. M., e. June 11th, '61; Granville co.; dt.

Hunt, J. P., e. June 11th, '61; Granville co.; dt.

Hunt, W. H., e. March 15th, '62; Granville co.; w. at Chancellorsville and Gettysburg; m. in action May 9th, '64.

Johnston, James R., e. June 11th, '61; Granville co.; w. at Cold Harbor; k. Sept. 17th, '62, at Sharpsburg.

Johnston, T. J., e. March 15th, '62; Granville co.

Johnston, Macon, e. March 15th, '62; Granville co.; dg. for disability.

Jenkins, S. D., e. July 8th, '62; Granville co.

Jones, Jeremiah, e. March 15th, '62; Granville co.

Jones, Wm. E., e. July 8th, '62; Granville co.; k. Sept. 17th, '62, at Sharpsburg.

Kavanaugh, John, e. June 11th, '61; Granville co.; w. May 31st, '62, at Seven Pines.

King, Jeremiah, e. March 15th, '62; Granville co.

Lancaster, Micajah, e. June 11th, '61; Granville co.

Lancaster, Washington, e. June 11th, '61; Granville co.; dg. April 7th, '63.

Lancaster, Joseph, e. June 11th, '61; Granville co.; d. September 25th, '61.

Lancaster, James, e. June 11th, '61; Granville co.; dg. for disability.

Lloyd, Jas. A., e. March 15th, '62; Granville co.; k. May 31st, '62, at Seven Pines.

May, T. H., e. June 11th, '63; Granville co.; tr.

Montague, L. J., e. July 8th, '62; Granville co.; d. January 16th, '63, in Va.

Maynard, J. P., e. July 8th, '62; Granville co.; w. and pr. at South Mountain; dg. Oct. 22d, '64.

O'Brian, Doctor, e. March 15th, '62; Granville co.; d. March 9th, '62.

Orrell, J. N., e. June 11th, '61; Granville co.

Overton, J. R., e. July 8th, '62; Granville co.; d. Dec. 5th, '62.

Peace, Henry, e. March 15th, '62; Granville co.; d.

Proctor, Richard, e. June 11th, '61; Orange co.; d. in Va.

Parham, A. C., e. June 11th, '61; Granville co.; dg. for disability.

Powell, Peter, e. June 11th, '61; Orange co.; dg. for disability.

Powell, A. S., e. June 11th, '61; Orange co.

Patterson, C. G., e. July 8th, '62; Granville co.

Parham, S. A., e. July 8th, '62; Granville co.; d. in North Carolina.

Reavis, S. W., e. June 11th, '61; Granville co.; c. September 19th, '64.

Reavis, T. C., e. July 8th, '62; Granville co.; dt.

Renn, Lewis, e. June 11th, '61; Granville co.

Roberts, W. H., e. June 11th, '61; Granville co.

Roberts, Thomas, e. March 15th, '02; Granville co.; d.

Robinson, P. A., e. June 11th, 61; Granville co.

Robinson. Lewis, e. June 11th, '61; Granville co.; w. May 31st, '62, at Seven Pines.

Robinson, George C., e. June 11th, '61; Granville co.; w. May 31st, '62, at Seven Pines.

Robinson, Alexander, e. June 11th, '61; Granville co.; pr. September 19th, '64.

Robinson, N. Y., e. June 11th, '61; Granville co.

Robinson, Ellis, e. March 15th, '62; Granville co.

Robinson, T. R., e. March 15th, '62; Granville co.; d.

Robinson, Robert, e. March 15th, '62; Granville co.; dg. September 8th, '62, for disability.

Robinson, David, e. March 15th, '62; Granville co.; d.

Robinson, Hillmon, e. December 9th, '63; Wake co.; pr. September 1st, '64.

Riley, Patrick, e. December 7th, '62; Virginia; m. May 12th, '04.

Short, J. T., e. June 11th, '61; Granville co.

Short, J. S., e. November 29th, '63; Wake co.

Stanton, R. T., e. October 18th, '64; Wake co.

Stone, J. T., e. March 15th, '62; Granville co.; w. at Seven Pines and Gettysburg.

Stewart, J. L., e. June 11th, '61; Granville co.; d. February, '62.

Staunton, G. G., e. July 8th, '62; Granville co.; w. July 20th, '04.

Satterwhite, J. F., e. July 8th, '62; Granville co.

Stewart, J. R., e. October 25th, '62; Granville co.; k. July, '63, at Gettysburg.

Tunstall, J. B., e. June 11th, '61; Granville co.; w. May 9th, '64.

Thompson, J. M., e. July 8th, '62; Granville co.; w. at Sharpsburg; dt.

Thompson, T. M., e. July 8th, '62; Granville co.

Turner, R. A., e. July 8th, '62; Granville co.

Upchurch, J. H., e. July 8th, '62; Granville co.; d. November 5th, '62.

Vaughan, Fielding, e. June 11th, '61; Granville co.; w. at Gettysburg; pr. Sept. 19th, '64.

Vaughan, Alexander, e. June 11th, '61; Granville co.; d. in Virginia.

Watkins, R. K., e. June 11th, '61; Granville co.

Wortham, J. W., e. June 11th, '61; Granville co.; d. June 15th, '63.

Wright, E. G., e. June 11th, '61; Granville co.; w. at Seven Pines; dt.

Williams, J. D., e. July 8th, '62; Granville co.

Weaver, W. H., e. July 8th, '62; Granville co.

Weaver, J. E., e. July 8th, '62; dt.

COMPANY H.

OFFICERS.

E. M. Faires, Captain, cm. June 12th, '61; Gaston co.; r. Dec. 1st, '61.

W. P. Hill, Captain, cm. Dec. 1st, '61; Gaston co.; p. from Sergeant.

H. G. Turner, Captain, cm. August 18th, '62; Granville co.; p. from ranks of Savannah Guards; w. and pr. July 1st, '63, at Gettysburg.

R. M. Ratchford, 1st Lieut., cm. June 12th, '61; Gaston co.; r. Dec., '61.

Joseph J. Wilson, 1st Lieut., cm. Dec., '61; Gaston co.; p. from Sergeant.

I. E. Hill, 2nd Lieut., cm. May 10th, '61; Gaston co.; p. from ranks.

Joseph B. F. Riddle, 1st Lieut., cm. May 10th, '62; Gaston co.; w. Sept. 30th, '64; p. from Sergeant.

T. N. Craig, 2d Lieut., cm. June 12th, '61; Gaston co.

J. M. Kendrick, 2d Lieut., cm. June 12th, '61; Gaston co.; pr. July 1st, '63, at Gettysburg.

W. S. Floyd, 2d Lieut., cm. ——; Gaston co.

NON-COMMISSIONED OFFICERS.

Joseph J. Wilson, 1st Sergeant, e. June 12th, '61; Gaston co.; p. 1st Lieutenant Dec., '61.

W. P. Hill, 2d Sergeant, e. June 12th, '61; Gaston co.; p. Captain December, '61.

Joseph B. F. Riddle, 3d Sergeant, e. June 12th, '61; Gaston co.; p. Lieut. May 10th, '62, and w. Sept., '64.

Wm. E. Wilson, 4th Sergeant, e. June 12th, '61; Gaston co.

J. N. Gulick, 5th Sergeant, e. June 12th, '61; Gaston co.; d. April 30th, '62, at Charlottesville.

J. E. Linebarger, 1st Corporal, e. June 12th, '61; Gaston co.; d. May 6th, '63, of w. received at Chancellorsville.

Robert N. Glenn, 2d Corporal, e. June 12th, '61; Gaston co.; k. May 3d, '63, at Chancellorsville.

J. F. Wilson, 3d Corporal, e. June 12th, '61; Gaston co.; p. Sergeant.

R. Featherston, 4th Corporal, e. June 12th, '61; Gaston co.; d. July 29th, '61, at Garysburg.

PRIVATES.

Adams, J. C., e. Aug. 1st, '62; Wilkes co.; k. July, '63, at Gettysburg.

Bailey, B., e. Aug. 1st, '62; w. May 12th, '64.

Barnes, J., e. Aug. 29th, '62; Surry co.

Berry, R. W., e. February 26th, '63; Gaston co.

Brison, J. H., e. April, '63; Gaston co.; w at Chancellorsville.

Berry, E. M., e. June 12th, '61; Gaston co.; dg. Nov. 10th, '61, for disability.

Beatty, F. M., e. June 12th, '61; Gaston co.; d. Aug. 6th, '62.

Bard, T. J., e. June 12th, '61; Gaston co.; d. January 1st, '62.

Bell, M., e. June 12th, '61; Gaston co.; dg. December 28th, '61.

Brison, J. B. F., e. June 12th, '61; Gaston co.

Camp, H. P., e. June 12th, '61; Cleveland co.; d. August, '61, in North Carolina.

Compton, M. V., e. June 12th, '61; Gaston co.; d. December 10th, '61.

Clinton, T. L., e. June 12th, '61; Gaston co.; p. Corporal; w. at Chancellorsville.

Connor, L. B., e. June 12th, '61; Gaston co.; tr. to Com. D.

Carson, J. H., e. Sept. 13th, '62; Alexander co.

Caudle, F. M., e. Aug. 1st, '62; Alexander co.

Craig, J. M., e. Sept. 23d, '61; Gaston co.; dg. Aug. 3d, '62.

Craig, T. V., e. December 28th, '62.

Damnon, J. T. R., e. September 3d, '63; w. at Chancellorsville.

Dancey, A. M., e. Aug. 1st, '62; Alexander co.

Durmine, J. B., e. Sept. 3d, '62; Alexander co.

Durmine, N. W., e. Sept. 3d, '62; Alexander co.; pr.

Eller, Eli, e. Sept. 3d, '62; Rowan co.; d. Dec., '62, at Richmond.

Eller, James, e. Sept. 3d, '62.

Eller, Samuel, e. Sept. 3d, '62; w. at Gettysburg.

Ford, E. M., e. June 12th, '61; Gaston co.; dg. September 29th, '62.

Ford, D. W., e. June 12th, '61; Gaston co ; k. July 1st, '63, at Gettysburg.

Ford, L. A., e. June 12th, '61; Gaston co.

Ford, J. C., e. June 12th, '61; Gaston co.; k. May, '63, at Chancellorsville.

Fite, W. F., e. June 12th, '61; Gaston co.; w. at Seven Pines; pr. May 12th, '64.

Floyd, W. S., e. June 12th, '61; Gaston co.; p. 2d Lieut.

Freeman, E. S., e. June 12th, '61; Gaston co.; w. at Sharpsburg.

Fronebarger, J. C., e. June 12th, '61; Gaston co.; w. at Sharpsburg and Richmond.

Faires, J. A., e. June 12th, '61; Gaston co.; dg for disability.

Frick, John, e. Sept. 4th, '62; k. July, '63, at Gettysburg.

Ford, J. A., e. February 26th, '63; Gaston co.

Fronebarger, J. C., e. June 12th, '61; Gaston co.; p. 1st Sergeant.

Grier, W. A., e. Sept. 1st, '62; Ashe co.; c.

Germon, H., e. Aug. 1st, '62; Wilkes co.

Glenn, J. W., e. June 12th, '61; Gaston co.; w. at Seven Pines and Sharpsburg.

Glenn, J. F., e. June 12th, '61; Gaston co.; d. December, '61, at Petersburg.

Glenn, D. L., e. June 12th, '61; Gaston co.; p. Corporal.

Groves, M. F., e. June 12th, '61; Gaston co.; dg. January 31st, '63, for w.

Groves, J. J., e. June 12th, '61; Gaston co.; d. June, '62, of w. received at Seven Pines.

Hill, J. E., e. June 12th, '61; Gaston co.; p. 2d Lieut., May 10th, '61.

Hucks, J. K., e. June 12th, '61; Gaston co.; dt.

Herron, D. P. F., e. June 12th, '61; Gaston co.; w. at Williamsburg.

Humphries, R. Z., e. June 12th, '61; Gaston co ; w. at Seven Pines.

Hampton, W., e. Aug. 1st, '62.

Hill, M., e. Aug. 1st, '62; w.

Hines, E. M., e. Sept. 3d, '62; dt.

Jarrett, G. W., e. June 12th, '61; Gaston co.; w. at Gettysburg.

Jenkins, J. L., e. June 12th, '61; Gaston co.; d. Aug. 25th, '61, in Va.

Johnston, T. H., e. June 12th, '61; Gaston co.

Johnston, A. J., e. June 18th, '61; Gaston co.; dg. Sept. 30th, '62.

Johnston, B. S., e. Aug. 1st, '62; Davie co.

Jarrett, C. F., e. Oct. 18th, '63; Gaston co.; pr.

Jay, J. H., e. June 12th, '61; Gaston co.; dg. May 12th, '63.

Kizer, J. C., e. June 12th, '61; Gaston co.; d. May 21st, '63, of w. received at Chancellorsville.

Kizer, E., e. June 12th, '61; Gaston co.; k. July 1st, '63, at Gettysburg; p. Corporal.

Long, D. L., e. June 12th, '61; Gaston co.; k. July 1st, '63, at Gettysburg.

Lenly, D. A., e. Sept. 4th, '62; Rowan co.

Lenly, B. T., e. September 4th, 62; Rowan co.

Longbottom, A., d. March 23d, '63.

Lyall, A., e. August 1st, 62; Ashe co.

Lyall, H., e. August 1st, '62; .Ashe co.

McCullough, J., e. June 12th, '61; Gaston co.; d. July 11th, '62, in Va.

McAllister, L. A., e. June 12th, '61; Gaston co.

Milling, W. R., e. June 12th. '61; Gaston co.; w. at Chancellorsville.

McLure, S. L., e. June 12th, '61; Gaston co.; d. July 5th, '63, of w. received at Gettysburg.

Mabery, F., e. September 3d, 62; Ashe co.

Manuel, G. E., e. September 3d, '62; Ashe co.; w. at Chancellorsville and Gettysburg.

Myers, C., Ashe co.; k. May 3d, '63, at Chancellorsville.

Millsaps, W. S., e. September 3d, '62; Ashe co.

Nesler, T. B., e. Sept. 7th, '62; Wilkes co.

O'Daniel, W. J., e. June 12th, '61; Gaston co.

Pearson, S. F., e. June 12th, '61; Gaston co.; d. April 6th, '62.

Parsons, J. H., Ashe co.

Paine, J. A. J., e. June 12th, '61; Gaston co.; w. at Malvern Hill; p. Sergeant.

Price, J. D. M., e. June 12th, '61; Gaston co.; d. June 10th, '63, of w. received at Chancellorsville.

Reid, R. H., e. June 12th, '61; Gaston co.; k. May, '63, at Chancellorsville.

Robinson, J. C., e. June 12th, '61; Gaston co.; d. June 4th, '62, of w. received at Seven Pines.

Rhodes, Christopher, e. June 12th, '61; Gaston co.; dg. July 20th, '62.

Ramsey, J., e. June 12th, '61; Gaston co.; c.

Reynolds, L., e. June 12th, '61; Gaston co.

Reid, W. E., e. August 1st, '62; Davie co.; w. at Chancellorsville.

Sarlott, W. F., k. July, '63, at Gettysburg.

Seaman, H. R., e. August 1st, '62; Davie co.

Stroud, R. S., e. Sepember 3d, '62.

Suryer, J. F., e. June 12th, '61; Gaston co.; dg. September 10th, '62.

Stowe, J. L., e. June 12th, '61; Gaston co.; p. Sergeant; d. August 2d, '63, of w. received at Gettysburg.

Stowe, H. W., e. June 12th, '61; Gaston co.; d. June 13th, '63, of w. received at Seven Pines.

Stowe, J. L., e. June 12th, '61; Gaston co.; p. Sergeant.

Smith, R. W., e. June 18th, '61; Gaston co.; d. March 16th, '63.

Smith, W. L., e. June 18th, '61; Gaston co.; w. at Chancellorsville.

Shannon, J. R., e. June 29th, '61; Gaston co.

Torrence, J. D., e. June 12th, '61; Gaston co.; w. at Sharpsburg and Chancellorsville and c.

Torrence, L. C., e. June 12th, '61; Gaston co.; dg. March 10th, '63.

Torrence, L., e. July 18th, '61; Gaston co.; p. Corporal; k. July 1st, '63, at Gettysburg.

Torrence, H. A., e. August 3d, '61; Gaston co.; p. Sergeant; w. at Chancellorsville.

Torrence, R. S., e. December 17th, '62; Gaston co.; w. at Chancellorsville.

Turner, H. G., e. Aug. 18th, '62; Granville co.; p. Captain; pr. July, '63.

Vandervort, W., e. September 4th, '62; Rowan co.; w. at Chancellorsville.

Venable, P., e. September 2d, '62; Rowan co.

Vandyke, L. S., e. June 12th, '61; Gaston co.; dt.

Warre, W. G., e. June 12th, '61; Gaston co.; dg. December 12th, '62.

Wallace, H. M., e. June 12th, '61; Gaston co.; d. of w. received at Gettysburg.

Wallace, J. T., e. July 18th, '61; Gaston co.; w. at Sharpsburg.

Wallace, L. J., e. July 18th, '61; Gaston co.

Wright, J. J., e. June 12th, '61; Gaston co.; w. at Seven Pines; k. June 27th, '62.

Wright, S., e. June 12th, '61; Gaston co.; d. January 8th, '62.

Wright, W. R., e. June 12th, '61; Gaston co.; dg. August 31st, '63, for w. at Sharpsburg.

Wright, W. H., e. June 12th, '61; Cleveland co.; dg. December 25th, '61.

Wright, Jasper, e. September 23d, '61; Gaston co.; d. August 9th, '62.

Wyatt, G. W., e. August 1st, '62; Rowan co.

Wyatt, J. E., e. September 4th, '62; Rowan co.; c. July 1st, '63.

Wyatt, W. W., e. September 4th, '62; Rowan co.; k. July, '63, at Gettysburg.

Wallace, D. S., e. Sept. 21st, '61; Gaston co.

Wilson, J. F., e. June 12th, '61; Gaston co.; pr. July 4th, '63.

— — —

COMPANY I.

OFFICERS.

Rufus Amis, Captain, cm. June 17th, '61; Granville co.

G. T. Baskerville, Captain, cm. May 10th, '62; Granville co.; p. from ranks.

G. B. Bullock, Captain, Granville co.; p. from 2d Lieutenant.

N. A. Gregory, 1st Lieut., cm. June 17th, '61; Granville co.

G. B. Bullock, 1st Lieut., Granville co.

J. D. Knott, 1st Lieut., cm. May 8th, '62; Granville co.; k. at Seven Pines.

A. M. Suria, 2d Lieut., cm. June 17th, '61; Georgia.

T. R. Carrington, 2d Lieut., cm. June 17th, '61; Granville co.

G. B. Bullock, 2d Lieut.; Granville co.; p. from ranks of 12th Regiment.

J. D. Knott, 2d Lieut., cm. November 16th, '61; Granville co.; p.

G. T. Sanford, 2d Lieut., cm. May 20th, '62; Granville co.

W. B. Sims, 2nd Lieut., cm. May 20th, '62; Granville co.; p. from ranks.

NON-COMMISSIONED OFFICERS.

R. D. Puryear, 1st Sergeant, e. June 17th, '61; Granville co.; dg. September 18th,
'61, for disability.

S. L. Puryear, 2d Sergeant, e. June 17th, '61; Granville co.; w. at Sharpsburg and
Chancellorsville.

H. Y. Demont, 3d Sergeant, e. June 17th, '61; Granville co.; d. September 18th, '61.

R. J. Hester, 4th Sergeant, e. June 17th, '61; Granville co.; w. at Seven Pines and
Chancellorsville.

W. J. Amis, 5th Sergeant, e. June 17th, '61; Granville co.; k. May 31st, '62, at
Seven Pines.

R. L. Watkins, 1st Corporal, e. June 17th, '61; Granville co.; dg. December, '61.

A. T. Puryear, 2d Corporal, e. June 17th, '61; Granville co.; d. September 23d. '61.

T. B. Morgan, 3d Corporal, e. June 17th. '61; Granville co.; k. July 1st, '63, at Gettysburg.

J. R. Blanks, 4th Corporal, e. June 17th, '61; Granville co.; dg. February 8th, '62.

PRIVATES.

Adcock, Robert, e. June 17th, '61; Granville co.; k. May, '63, at Chancellorsville.

Blackwell, James, e. June 17th, '61; Granville co.; pr. July 1st, '63.

Blackwell, Robert, e. June 17th, '61; Granville co.; k. July, '63, at Gettysburg.

Beal, J. W., e. June 17th, '61; Granville co.

Berkley, J. B., e. June 17th, '61; Granville co.; d. May 17th, '61.

Beasley, W. T., e. June 17th, '61; Granville co.; d. October 20th, '62.

Buchanan, R. S., e. June 17th, '61; Granville co.; dg. May 1st, '62.

Bailey, H. J., e. June 17th, '61; Granville co.

Briggs, H. T., e. June 17th, '61; Granville co.; k. at Sharpsburg.

Bullock, G. B., e. April 21st, '61; Warren co.; p. Lieut. from 12th Regiment, and then Captain.

Baskerville, G. T., e. April 15th, '61; Granville co.; p. Captain.

Beasley, S. H., e. June 17th, '61; Granville co.; p. Corporal; w. at Chancellorsville.

Chandler, R. R., e. June 17th, '61; Granville co.; p. Sergeant; c.

Chandler, D. W., e. June 17th, '61; Granville co.

Clark, J. L., e. June 17th, '61; Granville co.; k. near Kelley's Ford, Va.

Clark, James, e. July 8th, '62; Granville co.

Clark, A. S., e. July 8th, '62; Granville co.; w. at Sharpsburg.

Currin, David, e. June 17th, '61; Granville co.; d. October 1st, '61.

Currin, George, e. July 8th, '62; Granville co.; tr. to 55th Regiment.

Currin, William, e. March 1st, '62; Granville co.; pr.

Daniel, George, e. June 17th, '61; Granville co.

Daniel, William, e. June 17th, '61; Granville co.; d. in '61.

Duncan, J. K., e. July 8th, '62; Person co.; pr.

Duncan, David, e. July 8th, '62; Granville co.

Dixon, F. M., e. July 8th, '62; Granville co.

Eaks, Mark, e. June 17th, '61; Granville co.; k. May 31st, '62, at Seven Pines.

Eaks, Peter, e. February 28th, '62; Granville co.; d. March 15th, '63.

Eaks, Alexander, e. June 17th, '61; Granville co.; d. March, '61.

Eaks, Matthew, e. June 17th, '61; Granville co.; w. at Gettysburg.

Eaks, Mangum, e. March 1st, '62; Granville co.; d. June 9th, '62.

Faucett, George, e. June 17th, '61; Granville co.; k. at Malvern Hill.

Faucett, J. H., e. June 17th, '61; Granville co.

Faucett, William, e. June 17th, '61; Granville co.; w. at Seven Pines and Chancellorsville.

Frazier, J. S., e. June 17th, '61; Granville co.

Frazier, Thomas, e. July 8th, '62; Granville co.; w. at Gettysburg.

Gunn, A. L., e. June 17th, '61; Granville co.

Gunn, J. T., e. March 1st, '62; Granville co.; d. May 6th, '62.

Gunn, H., e. June 17th, '61; Granville co.; dg. February, '62, for disability.

Guy, William, e. June 17th, '61; Granville co.; d. December, '62.

Goodwyn, T. A., e. September 7th, '63.

Hester, H. J., e. June 17th, '61.

Hobgood, A., e. June 17th, '61; Granville co.; p. Sergeant; w. at Seven Pines.

Humphries, B. R., e. June 17th, '61; Granville co.; dg. February 14th, '62.

Hester, Robert, e. June 17th, '61; Granville co.; w. at Seven Pines.

Hart, William, e. June 17th, '61; Granville co.

Hart, J. W., e. February 28th, '62; Granville co.; p. Sergeant.

Jones, Robert, e. July 8th, '62; Granville co.; p. Corporal and pr.

Jones, J. A., e. February 28th, '62; Virginia.

Kinton, W. T., e. June 17th, '61; Granville co.

Knott, Robert, e. February 15th, '63; Granville co.

Knott, J. D., e. June 17th, '61; Granville co.; p. 2d Lieutenant; k. at Seven Pines.

Lloyd, Louis, e. June 17th, '61; Granville co.; dg. October 17th, '62.

Lloyd, David, e. July 8th, '62; Granville co.; k. July, '63, at Gettysburg.

Land, J. B., e. June 17th, '61; Granville co.; k. May 31st, '62, at Seven Pines.

Lynch, Dennis, e. July 8th, '62; Granville co.; w. at Gettysburg.

34   **NORTH CAROLINA - 23rd REGIMENT INFANTRY - Roster and History**

****

Mulchi, G. W., e. Feb. 28th, '62; Granville.

Norwood, J. B., e. June 17th, '61; Granville co.

Newton, S. L., e. Feb. 28th, '62; Granville co.

Newton, G. W., e. Feb. 28th, '62; Granville co.

Newton, Isaac, e. Feb. 28th, '62; Granville co.; d. March 18th, '82.

Norman, Thomas, e. July 8th, '62; Granville co.; d.

O'Brian, R. S., e. June 17th, '61; Granville co.; k. at Seven Pines.

O'Brian, Wm. H., e. June 17th, '61; Granville co.; w. at Chancellorsville.

Overby, James, e. June 17th, '61; Granville co.

Overby, Thomas, e. June 17th, '61; Granville co.; dg. August 10th, '63.

Overby, David, e. February 28th, '62; Granville co.

Puryear, W. D., e. June 17th, '61; Granville co.; d. May 24th, '63.

Puryear, H. T., e. June 17th, '61; Granville co.; w. at Chancellorsville.

Puryear, J. R., e. June 17th, '61; Granville co.

Puryear, Ruffin, e. June 17th, '61; Granville co.; dg. Nov. 18th, '61.

Parham, S. R., e. June 17th, '61; Granville co.

Puckett, J. T., e. June 17th, '61; Granville co.; d. Dec. 14th, '61.

Royster, Willis, e. July 8th, '62; Granville co.

Royster, M. D., e. June 17th, '61; Granville co.; d. June 8th, '62.

Robinson, Banister, e. July 8th, '62; Granville co.; pr.

Sadler, John, e. June 17th, '61; Granville co.

Sadler, Robert, e. February 28th, '62; Granville co.

Sandford, J. T., e. June 17th, '61; Granville co.; k. July, '63, at Gettysburg.

Sims, W. B., e. February 28th, '62; Granville co.; p. 2d Lieut. May 20th, '62.

Stone, R. H., e. June 17th, '61; Granville co.; w. at Seven Pines; p. Sergeant.

Tillotson, W. B., e. June 17th, '61; Granville co.; w. at Seven Pines.

Tillotson, R., e. June 17th, '61; Granville co.; k. May 31st, '62, at Seven Pines.

Tillotson, M., e. June 17th, '61; Granville co.; d. June 18th, '62.

Talley, James, e. June 17th, '61; Granville co.; d. April 4th, '62.

Tuck, W. A., e. March 1st, '62; Granville co.; dg. April 28th, '63.

Thomas, J. W., e. July 8th, '62; Granville co.

Taylor, A. P., e. November 27th, '64.

Tyach, J. L., e. May 20th, '61.

Wade, D. H., e. April 15th, '63; Virginia.

West, C. H., e. June 17th, '61; Granville co.

West, E. B., e. February 28th, '62; Granville co.

West, H., e. February 28th, '62; Granville co.

West, Stephen, e. February 28th, '62; Granville co.; k. May, '63, at Chancellorsville.

West, F., e. June 17th, '61; Granville co.

West, A., e. February 28th, '62; Granville co.; pr.

West, Smith, e. June 17th, '61; Granville co.; dg. May 12th, '61.

West, Robert, e. July 8th, '62; Granville co.; pr.

West, J. B., e. July 8th, '62; Granville co.

Wilbour, J. A., e. June 17th, '61; Granville co.; k. July, '63, at Gettysburg.

Wilkinson, W. D., e. February 28th, '62; Virginia; d. July 4th, '63, of w. received
at Gettysburg.

Wilkinson, J. D., e. July 8th, '62; Granville co.; dt.

Wilkinson, W. L., e. July 8th, '62; Granville co.; dt.

Wilson, S., e. July 8th, '62; Granville co.; pr. July 1st, '63.

Wilson, L. Peter, e. July 8th, '62; Granville co.; pr. September 19th, '64.

Wilson, Benjamin, e. June 17th, '61; Granville co.; d. January 8th, '62.

Willis, James, e. July 8th, 62; Granville co.; w. at South Mountain.

York, J. E., e. June 17th, '61; Granville co.; dg.

Yancey, Thomas, e. June 17th, '61; Granville co.; d. October 13th, '62.

NORTH CAROLINA - 23rd REGIMENT INFANTRY - Roster and History   35

****

COMPANY K.

OFFICERS.

Robert D. Johnston, Captain, cm. June 22d, '61; Lincoln co.; p. Lieut. Col. May 10th, '62, and Brigadier General in '62.

William H. Johnston, Captain, cm. May 10th, '62; Lincoln co.; p. from 1st Lieut.; k. July 1st, '63, at Gettysburg.

W. H. Johnston, 1st Lieut., cm. June 22d, '61; Lincoln co.; p. and k.

Daniel Reinhardt, 1st Lieut., cm. September, '62; Lincoln co.

John F. Goodson, 2d Lieut., cm. June 22d, '61; Lincoln co.

G. W. Hunter, 2d Lieut., cm. June 22d, '62; Lincoln co.

Daniel Reinhardt, 2d Lieut., cm. May 10th, '62; Lincoln co.

J. A. Caldwell, 2d Lieut., cm. September 6th, '62; Lincoln co.

William M. Munday, 2nd Lieut., cm. in September, '62; Lincoln co.; p. from ranks; w. at Malvern Hill.

H. W. Fullenwider, 2nd Lieut., cm. in May, '63; Lincoln co.; p. from ranks.

NON-COMMISSIONED OFFICERS.

C. L. Gattis, 1st Sergeant, e. June 22d, '61; Lincoln co.; dt. Commissary Sergeant.

S. C. Little, 2d Sergeant, e. June 22d, '61; Lincoln co.; d. October 9th, '61.

James T. Johnson, 3d Sergeant, e. June 22d, '61; Lincoln co.; p. Captain and Brigade Commissary.

James Little, 4th Sergeant, e. June 22d, '61; Lincoln co.; d. Dec. 18th, '61, in Va.

Samuel T. Thompson, 1st Corporal, e. June 22d, '61; Lincoln co.; p. Lieut. in 28th Regiment.

Spencer P. Shelton, 2d Corporal, e. June 22d, '61; Lincoln co.; d. Nov. 23d, '61.

William Bunch, 3d Corporal, e. June 22d, '61; Lincoln co.; d. Dec. 3d, '61.

Daniel Reinhardt, 4th Corporal, e. June 22d, '61; Lincoln co.; p. Lieutenant May 10th, '62.

PRIVATES.

Asberry, Thos. H., e. June 22d, '61; Lincoln co.; k. June 27th, '62, at Cold Harbor.

Allen, Henry H., e. August 20th, '61; Lincoln co.; w. at Gettysburg.

Allen, Josiah, e. August 20th, '61; Lincoln co.; d. at Fort Delaware.

Allen, W. H., e. August 20th, '61; Lincoln co.; d. March 13th, '63, in Virginia.

Ballard, Franklin M., e. June 22d, '61; Lincoln co.

Ballard, Marcus G., e. June 22d, '61; Lincoln co.

Benton, Thomas H., e. June 22d, '61; Lincoln co.; w. at South Mountain.

Blythe, C. N., e. June 22d, '61; Mecklenburg co.

Brotherton, W. H., e. August 20th, '62; Lincoln co.

Bunch, James W., e. June 22d, '61; Lincoln co.; w. July 1st, '62, at Malvern Hill; d. August 1st, '64.

Baker, Jehu, e. August 20th, '62; Lincoln co.; c. July 8th, '63.

Barnhill, Samuel, e. August 20th, '62; Lincoln co.; d. Dec. 23d, '62, in Virginia.

Bennick, H. D., e. August 20th, '62; Lincoln co.; d. October 26th, '62, in Virginia.

Caldwell, J. A., e. June 22d, '61; Lincoln co.; p. Lieut. September, '62.

Conner, H. W., e. June 22d, '61; Lincoln co.; w. May 31st, '62, at Seven Pines; dg.

Carpenter, Lawson, e. August 20th, '62; Lincoln co.; d. Dec. 1st, '62, in Virginia.

Dillinger, George S., e. June 22d, '61; Lincoln co.; d. of w. at Gettysburg.

Dougherty, Henry, e. August 21st, '62; Lincoln co.; d. at Richmond.

Engle, David, e. Aug. 21st, '62; Lincoln co.; w. May 3d, '63, at Chancellorsville; dt.

Fullenwider, H. W., e. February 21st, '62; Mecklenburg co.; w. Sept. 1st, '62, at Sharpsburg; p. Lieut. May, '63.

Gabriel, Abram A., e. June 22d, '61; Catawba co.

Gabriel, A. Alonzo, e. June 22d, '61; Catawba co.; m. in action July 1st, '63.

Gabriel, Monroe M., e. June 22d, '61; Catawba co.; w. at Malvern Hill July 1st, '63.

Gilbert, Robert, e. August 20th, '62; Lincoln co.; w. at Gettysburg.

Goodson, Albert A., e. June 22d, '61; Lincoln co.

Guthrie, Columbus, e. August 20th, '62; Lincoln co.; w. at Chancellorsville.

Gattis, C. L., e. June 22d, '61; Lincoln co.

Hager, Green A., e. June 22d, '61; Mecklenburg co.; pr. July 1st, '63.

Hager, Robert T., e. June 22d, '61; Lincoln co.

Hager, Sidney, e. June 22d, '61; Mecklenburg co.; w. at Seven Pines and Gettysburg.

Hager, Monroe L., e. August 20th, '62; Lincoln co.; d. Oct. 27th, '62, in Va.

Hager, Brown, e. August 20th, '62; Lincoln co.; k. July 1st, '63, at Gettysburg.

Hager, O. W., e. November 10th, '63; d. April 20th, '64.

Hall, Andrew J., e. June 22d, '61; Mecklenburg co.; dt.

Hall, Thomas C., e. June 22d, '61; Mecklenburg co.

Henderson, Thomas C., e. June 22d, '61; Mecklenburg co.; w. at Chancellorsville and m. Sept. 19th, '64.

Harris, R. T., e. August 20th, '62; Lincoln co ; w. at Chancellorsville.

Howard, H. H., e. June 22d, '61; Caldwell co.; d. Oct. 1st, '64.

Howard, Jackson, e. August 20th, '62; Lincoln co.; w. at Chancellorsville.

Howard, W. G., e. August 20th, '62; Lincoln co.

Hoyle, Alfred E., e. June 22d, '61; Lincoln co.; k. May 31st, '62, at Seven Pines.

Hunter, H. S., e. June 22d, '61; Lincoln co.; w. at Seven Pines.

Hoover, Alexander, e. August 20th, '62; Lincoln co.; w. at Chancellorsville and d. August 1st, '64.

Hoover, Edney, e. August 20th, '62; Lincoln co.

Hoover, Wesley, e. August 20th, '62; Lincoln co.; d. at Richmond.

Johnson, James H., e. June 22d, '61; Mecklenburg co.; w. at Seven Pines and m. in action July 1st, '63.

Killiam, Jacob F., e. June 22d, '61; Lincoln co.; w. twice and k. May 3d, '63, at Chancellorsville.

Killiam, John A., e. June 22d, '61; Lincoln co.; pr. and d. at Point Lookout.

King, James G., e. June 22d, '61; Lincoln co.; dg. in October, '62.

King, Wm. O., e. August 20th, '62; Lincoln co.; d. April 19th, '63, in Va.

Linebarger, Jacob, e. June 22d, '61; Catawba co.; tr. to 28th Regiment.

Little, Alfred B., e. June 22d, '61; Lincoln co.; d. April 5th, '62, at Richmond.

Little, Robert W., e. June 22d, '61; Mecklenburg co.; w. at Chancellorsville.

Little, Hugh, e. June 22d, '61; Lincoln co.; w. twice; k. July 1st, '63, at Gettysburg.

Little, James, e. June 22d, '61; Lincoln co.; w. at Malvern Hill; k. May 3d, '63, at Chancellorsville.

Little, James B., e. August 20th, '62; Catawba co.; d. at Richmond.

Little, George W., e. June 20th, '61; Catawba co.; d. at Mount Jackson.

Little, Robert B., e. August 30th, '62; Lincoln co.

Little, John B., e. August 30th, '62; Lincoln co.; w. at Chancellorsville; dt.

Lockman, Elisha P., e. June 22d, '61; Lincoln co.; w. at Gettysburg.

Lockman, Levi A., e. June 22d, '61; Lincoln co.

Lockman, William L., e. June 22d, '61; Lincoln co.; w. at Seven Pines.

Long, Jacob L., e. June 22d, '61; Lincoln co.

Long, John A., e. August 20th, '62; Lincoln co.

Lynch, Isaac, e. June 22d, '61; Lincoln co.; k. May, '62, at Seven Pines.

Lingerfelt, Daniel, e. August 20th, '62; Lincoln co.; d. November 23d, '62, at Strasburg.

Maberry, Josiah, Alexander co.

Maberry, Noah, e. August 20th, '62; Alexander co.; d. at Winchester June 1st, '64.

McCaul, Joseph, e. June 22d, '61; Catawba co.; d. April 10th, '62, at Richmond.

Munday, Josiah F., e. June 22d, '61; Catawba co.; w. at Chancellorsville; dg. December 21st, '64.

Munday, Marcus, e. June 22d, '61; Lincoln co.; w. at Seven Pines.

Munday, William M., e. June 22d, '61; Lincoln co.; w. at Malvern Hill; p. Lieutenant September, '62.

Moore, James E. C., e. June 22d, '61; Gaston co.; k. May 31st, '62, at Seven Pines.

Mallis, James M., e. August 20th, '62; Catawba co.; k. July 1st, '63, at Gettysburg.

Natz, J. F., e. August 20th, '62; Iredell co.; d. August 15th, '64.

Natz, J. H., e. June 22d, '61; Iredell co.; dt.

Nance, Joseph C., e. June 22d, '61; Lincoln co.

Nance, Marcus, e. June 22d, '61; Lincoln co.; d. of w. received at Seven Pines.

Nance, John F., e. August 20th, '62; Lincoln co.

Nixon, James, e. August 20th, '62; Lincoln co.

Proctor, Thomas H., e. June 22d, '61; Lincoln co.; w. at Gettysburg.

Proctor, Richard G., e. August 20th, '62; Lincoln co; dt.

Pleasant, B. N., e. November 10th, '63; Wake co.

Regan, John A., e. June 22d, '61; Lincoln co.; d. at Richmond.

Randleman, Pink L., e. June 22d, '61; Lincoln co.; w. at Gettysburg.

Robinson, Samuel, e. June 22d, '61; Lincoln co.

Robertson, R., e. June 22d, '61; Lincoln co.

Rodgers, Erasmus, e. June 22d, '61; Mecklenburg co.; k. May 31st, '62, at Seven Pines.

Ruddock, Wm. O., e. June 22d, '62; Mecklenburg co.; w. at Seven Pines and dt.

Rankin, John N., e. August 20th, '62; Lincoln co.; d. October 25th, '62, at Winchester.

Ross, Obed, e. August 20th, '62; Lincoln co.; d. March 1st, '64.

Russ, Jonas, e. August 20th, '62; Lincoln co.

Rudisill, Philip, e. August 20th, '62; Lincoln co.; d. in Va.

Shelton, Albert, e. June 22d, '61; Lincoln co.; k. at South Mountain.

Shelton, Thomas, e. June 22d, '61; Lincoln co.; m. in action July 1st, '63.

Smith, Ephraim, e. August 20th, '62; Lincoln co.; w. at Chancellorsville.

Thompson, John H., e. June 22d, '61; Lincoln co.; w. at Chancellorsville.

Turbyfield, Francis, e. June 22d, '61; Catawba co.; w. at Gettysburg.

Thompson, John T., e. August 20th, '62; Lincoln co.; k. July 1st, '63, at Gettysburg.

Washam, John R., e. June 22d, '61; Mecklenburg co.; dt.

White, John, e. June 22d, '61; Mecklenburg co.; w. at Gettysburg.

Willis, Josiah, e. August 20th, '62; Lincoln co.

Womack, Elisha, e. August 20th, '62; Lincoln co.; w. at Chancellorsville.

Workman, Jacob, e. August 20th, '62; Lincoln co.

Yount, D., e. August 20th, '62; Lincoln co.

Yount, E. M., e. August 20th, '62; Iredell co.; w. Sept. 19th, '64.

TWENTY-THIRD REGIMENT.

1. J. H. Horner, Captain, Co. E.   4. V. E. Turner, Captain, Quart. Master.
2. Frank Bennett, Captain, Co. A.   5. Abner D Peace, Captain, Co. E.
3. H. G. Turner, Captain, Co. H.   6. Geo T. Baskerville, Captain, Co. I.
                    7. Jas. A. Breedlove, Captain, Co. G.

TWENTY-THIRD REGIMENT.

BY

CAPTAIN V. E. TURNER, A. Q. M.

H. C. WALL, SERGEANT COMPANY A.

Up to the re-arrangement of the regimental numbers following the Confederate Conscription Act, which went into effect 17 May, 1862, this regiment had been known as the Thirteenth Regiment of North Carolina Volunteers. The reason of the change is very clearly given by Major Gordon in the history of the organization. As repetition is, as far as possible, to be avoided in these sketches we will not give it here.

No North Carolinians were more forward in the cause of Southern defence than the men who formed the Twenty-third. They were among the first to respond when the State called upon her sons to repel invasion. The organization of most, if not all the companies, ante-date the Ordinance of Secession, passed 20 May, 1861.

This was only ten days after the act authorizing their enlistment was passed. Of course in this case, as in many others, the action of the State had been foreseen and anticipated, and the raising of companies had begun before.

The act authorizing the enlistment of the ten regiments of "State Troops" had been passed on 8 May, two days earlier.

The power of appointing all commissioned officers in the "State Troops" was lodged in the Governor. But the "Volunteers" to which the Twenty-third, then the Thirteenth, belonged, were empowered to elect their own officers, to be commissioned by the Governor. The men of each company were to elect their respective Line or Company Officers. The Line Officers were, by balloting among themselves, to elect Field or Regimental officers. The enlistment for the "Volunteers" was for twelve months; that of the "State Troops" as long as the war lasted. It is hardly necessary to add that both of the above classes of troops were in fact volunteers, the enlistment of both being entirely voluntary.

The personnel of the Twenty-third was doubtless as representative of the diverse racial strains of the State as any command raised within her borders. The three companies raised in Granville County, were virtually pure English, descendants of the early Virginia settlers who later settled in this State. In the company from Richmond and Anson Counties there was a strong infusion of Highland Scotch, descendants of the stout-hearted, strong armed Culloden lads who were "out wi' Charlie in the '45." In those from Catawba, Lincoln and Gaston, the German stock, that trending down from Pennsylvania had largely settled that part of the State, abounded. While the names in these and other companies from that region show the presence of many Scotch-Irish who had been co-settlers with the Germans.

The regiment was composed of the following companies. We give the original name which each company bore, and the county in which it was raised. Seeking to do justice to all, we give as complete as we are able to make it, a roster of the Line and Field officers, showing the promotions and casualties to the end of the war. We regret that lack of space excludes that of equally worthy non-commissioned officers and privates. But North Carolina has not been unmindful of them. All and the casualties of each, though not as accurately as could be wished, down to the humblest, appear in the general roster of which a large number of copies were published by the State in 1882.

COMPANY A—*Anson Ellis Rifles, Anson County*—Captain Wm. F. Harlee, of Anson County; commissioned May 22, 1861, resigned December 15, 1861. Captain James M. Wall, of Anson County, commissioned December 15, 1861. Captain Frank Bennett, of Anson County, commissioned May 10, 1862; promoted from First Sergeant; wounded May 29, 1862; wounded at Chancellorsville; wounded May 12, 1864, at Spottsylvania Court House; wounded at Hatcher's Run. W. D. Redfearne, First Lieutenant, of Anson County; commissioned May 22, 1861. James C. Marshall, First

Lieutenant, of Anson County; commissioned May 10, 1862; transferred as Adjutant to Fourteenth Regiment in 1862. John M. Little, Second Lieutenant, of Anson County; commissioned May 22, 1861. James Crowder, Second Lieutenant, of Anson County; commissioned May 22, 1861; wounded and captured at Sharpsburg; wounded at Lynchburg June, 1864. Samuel F. Wright, Second Lieutenant, of Anson County; commissioned May 10, 1862; captured at Gettysburg.

COMPANY B—*Hog Hill Guards, Lincoln County*—Geo. W. Seagle, Captain, Lincoln County; commissioned May 23, 1861. Wesley Hadspeth, Captain, Lincoln County; commissioned May 10, 1862; promoted from ranks; wounded at Sharpsburg; killed at Chancellorsville May 3, 1863. G. W. Hunter, Captain, Lincoln County; promoted from ranks. Josiah Holbrook, Captain, Lincoln County; promoted from ranks. T. J. Seagle, First Lieutenant, Lincoln County; commissioned May 23, 1861. M. H. Shuford, First Lieutenant, Lincoln County; commissioned May 23, 1861. Lee Johnson, Second Lieutenant, Lincoln County; commissioned May 23, 1861. S. A. Shuford, Second Lieutenant, Lincoln County; commissioned May 10, 1862. Wm. R. Sloan, Second Lieutenant, Mecklenburg County; commissioned May 10, 1862. M. H. Shuford, Second Lieutenant, Lincoln County; commissioned May 10, 1862. W. A. Thompson, Second Lieutenant, Lincoln County; commissioned May 10, 1862. M. M. Hines, Second Lieutenant, Lincoln County; commissioned November 20, 1861; prisoner September 19, 1864.

COMPANY C—*Montgomery Volunteers No. 1*—C. J. Cochrane, Captain, of Montgomery County; commissioned May 27, 1861. E. J. Christian, Captain, of Montgomery County, commissioned May 10, 1862; promoted Major May 10, 1862, and killed May 31, 1862 at Seven Pines. A. F. Scarborough, Captain, of Montgomery County; commissioned May 10, 1862; killed May 30, 1862. E. H. Lyon, Captain, of Granville County; commissioned May 31, 1862; transferred from Company E; prisoner September 19, 1864. E. J. Christian, First Lieutenant, of Montgomery County;

commissioned May 27, 1861; promoted and killed. John R. Nicholson, First Lieutenant, of Montgomery County; commissioned May 10, 1862. E. J. Garris, Second Lieutenant, of Montgomery County; commissioned May 10, 1862; killed W. Montgomery, Second Lieutenant, of Montgomery County; commissioned May 27, 1861. Jeremiah Coggins, Second Lieutenant, of Montgomery County; commissioned May 10, 1862; prisoner at Gettysburg July 1, 1863; one of the 600 officers placed under Confederate fire at Charleston, S. C.; died at Fort Delaware. A. F. Saunders, Second Lieutenant, of Montgomery County; commissioned May 10, 1862; killed at Spottsylvania May 9, 1864. J. P. Leach, Second Lieutenant, of Montgomery County; commissioned April 14, 1863.

COMPANY D—*Pee Dee Guards*—Lewis H. Webb, Captain, of Richmond County; commissioned May 30, 1861; resigned. A. T. Cole, Captain, of Richmond County; commissioned May 10, 1862; wounded at Sharpsburg; wounded and captured at Chancellorsville; captured at Spottsylvania C. H. May 12, 1864; one of the 600 officers placed under Confederate guns at Charleston, S. C. James S. Knight, First Lieutenant, of Richmond County; commissioned May 30, 1861; killed at Chancellorsville May 3, 1863. Risden T. Nichols, First Lieutenant, of Richmond County; commissioned May 10, 1862; died in 1862. J. H. Chappell, First Lieutenant, of Richmond County. John W. Cole, Second Lieutenant, of Richmond County; commissioned May 30, 1861. B. H. Covington, Second Lieutenant, of Richmond County; commissioned May 30, 1861. W. C. Wall, Second Lieutenant, of Richmond County; commissioned October 17, 1861; promoted Captain Company F; wounded at Monacacy July 1864. James H. Chappell, Second Lieutenant, of Richmond County; commissioned October 10, 1862; severely wounded at Chancellorsville; captured. E. A. McDonald, Second Lieutenant, of Richmond County; commissioned October 10, 1862; severely wounded at Chancellorsville.

COMPANY E—*Granville Plough Boys, Granville County*— J. H. Horner, Captain, of Granville County; commissioned June 5, 1861. B. F. Bullock, Captain, of Granville County; commissioned ————. E. E. Lyon, First Lieutenant, of

Granville County; commissioned June 5, 1861. T. W.
Moore, First Lieutenant, of Granville County; commissioned
August 15, 1861. J. H. Mitchell, Second Lieutenant, of
Granville County; commissioned June 5, 1861. A. D.
Peace, Second Lieutenant, of Granville County; commission-
ed June 5, 1861; wounded twice. R. V. Minor, Second
Lieutenant, of Granville County; commissioned September
25, 1862. E. H. Lyon, Second Lieutenant, of Granville
County; commissioned November 12, 1861; transferred as
Captain of Company C. B. F. Bullock, Second Lieutenant,
of Granville County; commissioned December 6, 1861. J.
T. Bullock, Second Lieutenant, of Granville County; com-
missioned May 10, 1862; captured May 12, 1864; one of
the 600 officers placed under Confederate guns at Charleston,
S. C. A. S. Webb, Second Lieutenant, of Granville County;
commissioned May 10, 1862; resigned.

COMPANY F—*Catawba Guards, Catawba County*—M. L.
McCorkle, Captain, of Catawba County; commissioned June
6, 1861. W. C. Wall, Captain, of Richmond County; commissioned May 10, 1864. Jacob H. Miller, First Lieutenant, of Catawba County; commissioned June 6, 1861. T.
W. Wilson, First Lieutenant, of Catawba County; killed at
Spottsylvania May 10, 1864. M. L. Helton, Second Lieutenant, of Catawba County; commissioned June 6, 1861.
R. A. Cobb, Second Lieutenant, of Catawba County; commissioned June 6, 1861. G. P. Clay, Second Lieutenant, of
Catawba County; commissioned May 10, 1862. T. W. Wilson, Second Lieutenant, of Catawba County; commissioned
May 10, 1862. W. C. Wall, Second Lieutenant, of Richmond County; commissioned May 10, 1862.

COMPANY G—*Granville Rifles*—C. C. Blacknall, Captain,
of Granville County; commissioned June 11, 1861; wounded
at Seven Pines; promoted Major May 31, 1862; captured at
Chancellorsville; wounded and captured at Gettysburg;
promoted Colonel August, 1863; mortally wounded
September 19, 1864. I. J. Young, Captain, of Granville
County; commissioned May 31, 1862; wounded May 31,
1862, at Seven Pines; resigned August 1862; wounded at
Malvern Hill. T. J. Crocker, Captain, of Granville County;

commissioned August 15, 1862; wounded, disabled and re-signed.   James A. Breedlove, Captain, of Granville County; commissioned in 1864; wounded; promoted from First Lieutenant.   Isaac J. Young, First Lieutenant, of Granville County; commissioned June 11, 1861; promoted, wounded, and resigned.   T. J. Crocker, First Lieutenant, of Granville County; commissioned May 31, 1862; promoted, wounded, and resigned; J. A. Breedlove, First Lieutenant, of Granville County; commissioned June 11, 1861; promoted and wounded.   Washington F. Overton, First Lieutenant, of Granville County; commissioned in 1864; wounded and burned in woods at Chancellorsville.   G. W. Kittrell, Second Lieutenant, of Granville County; commissioned June 11, 1861.   Vines E. Turner, Second Lieutenant, of Granville County; commissioned June 11, 1861; promoted Adjutant May 10, 1862; wounded at Cold Harbor June 27, 1862; promoted Assistant Quartermaster in 1863.   T. J. Crocker, Second Lieutenant, of Granville County; commissioned May 10, 1862; promoted.   William F. Gill, Second Lieutenant, of Franklin County; commissioned May 10, 1862; promoted from Sergeant-Major; killed at Malvern Hill.   W. F. Overton, Second Lieutenant, of Granville County; commissioned August 15, 1862; promoted and killed.   J. A. Breedlove, Second Lieutenant, of Granville County; commissioned August 15, 1862; promoted and wounded.   C. W. Champion, Second Lieutenant, of Granville County; commissioned November 1, 1862; killed at Gettysburg.

COMPANY H—*Gaston Guards*—E. M. Faires, Captain, of Gaston County; commissioned June 12, 1861; resigned December 1, 1861.   W. P. Hill, Captain, of Gaston County; commissioned December 1, 1861; promoted from Sergeant.   H. G. Turner, Captain, of Granville County; commissioned August 18, 1862; promoted from ranks of Savannah Guards; desperately wounded and captured July 1, 1862, at Gettysburg. R. M. Ratchford, First Lieutenant, of Gaston County; commissioned June 12, 1861; resigned December, 1861. Jos. J. Wilson, First Lieutenant, of Gaston County; commissioned December, 1861; promoted from Sergeant.   Joseph B. F. Riddle, First Lieutenant, of Gaston County; commissioned

May 10, 1862; wounded September 30, 1864; promoted from Sergeant. J. E. Hill, Second Lieutenant, of Gaston County; commissioned May 10, 1861; promoted from ranks. T. N. Craig, Second Lieutenant, of Gaston County; commissioned June 12, 1861. J. M. Kendrick, Second Lieutenant, of Gaston County; commissioned June 12, 1861; captured July 1, 1863, at Gettysburg. W. S. Floyd, Second Lieutenant, of Gaston County; commissioned ————.

COMPANY I—*Granville Stars*—Rufus Amis, Captain, of Granville County; commissioned June 17, 1861. G. T. Baskerville, Captain, of Granville County; commissioned 1863; killed at Gettysburg. G. B. Bullock, Captain, of Granville County; promoted from Second Lieutenant. N. A. Gregory, First Lieutenant, of Granville County; commissioned June 17, 1861; wounded and disabled at Chancellorsville. G. B. Bullock, First Lieutenant, of Granville County. J. D. Knott, First Lieutenant, of Granville County; commissioned May 8, 1862; killed at Seven Pines. A. M. Luria, Second Lieutenant, of Georgia; commissioned June 17, 1861; killed at Seven Pines. T. R. Carrington, Second Lieutenant, of Granville County; commissioned June 17, 1861. G. B. Bullock, Second Lieutenant, of Granville County; promoted from ranks of Twelfth Regiment. J. D. Knott, Second Lieutenant, of Granville County; commissioned November 16, 1861; promoted and killed. G. T. Sanford, Second Lieutenant, of Granville County; commissioned May 20, 1862. W. B. Sims, Second Lieutenant, of Granville County; commissioned May 20, 1862; promoted from ranks.

COMPANY K—*Beattie's Ford Riflemen, Lincoln County*— Robert D. Johnston, Captain, of Lincoln County; commissioned June 22, 1861; promoted Lieutenant-Colonel May 10, 1862, and Brigadier-General in 1863. William H. Johnston, Captain, of Lincoln County; commissioned May 10, 1862; promoted from First Lieutenant; captured July 1, 1863, at Gettysburg. W. H. Johnston, First Lieutenant, of Lincoln County; commissioned June 22, 1861; promoted and captured. Daniel Reinhardt, First Lieutenant, of Lincoln County; commissioned September, 1862. John F. Goodson, Second Lieutenant, of Lincoln County; commissioned June 22, 1861. G. W. Hunter, Second Lieutenant, of Lincoln County; commissioned June 22, 1861. Daniel Reinhardt, Second Lieutenant, of Lincoln County; commissioned May 10, 1862. J. A. Caldwell, Second Lieutenant, of Lincoln County; commissioned September 6, 1862. William M. Munday, Second Lieutenant, of Lincoln County; commissioned September, 1862; promoted from ranks; wounded at Malvern Hill. H. W. Fullenwider, Second Lieutenant, of Lincoln County; commissioned in May, 1863; promoted from ranks; killed.

Nine of these companies were assembled in camp near Weldon, N. C., and between that place and Garysburg, two miles distant, in June, 1861. Here the boys underwent a little more drilling than they liked. But they were patriots, one and all, and as some drilling might possibly be necessary even to whip Yankees, they submitted cheerfully. The other company, the Anson Ellis Rifles, remained in camp at Raleigh till ordered to join the regiment as it left for Virginia. Garysburg was the point of rendezvous. Here, in obedience to orders, the Line Officers of the ten companies met 10 July and elected Field Officers for the regiment as follows. The date, 10 July, 1861, shows the officers then elected. Other dates show the result of subsequent elections and promotions:

FIELD AND STAFF OFFICERS.

JOHN F. HOKE, Colonel, of Lincoln County; commissioned July 10, 1861.

DANIEL H. CHRISTIE, Colonel, of Granville County; commissioned May 10, 1862; wounded at Seven Pines; wounded at Cold Habor; mortally wounded July 1, 1863, at Gettysburg; died in Winchester August, 1863.

CHARLES C. BLACKNALL, Colonel, of Granville County; commissioned August 15, 1863; wounded at Seven Pines; captured at Chancellorsville; wounded and captured at Gettysburg; mortally wounded and captured at Winchester September 19, 1864; died November 6, 1864.

WM. S. DAVIS, Colonel, of Warren County; commissioned October 1864; transferred from Twelfth Regiment; wounded.

JOHN W. LEAK, Lieutenant-Colonel, of Richmond County; commissioned July 10, 1861.

ROBT. D. JOHNSTON, Lieutenant-Colonel, of Lincoln County; commissioned May 10, 1862; wounded at Seven Pines; wounded at Gettysburg; promoted Brigadier-General July, 1863; wounded at Spottsylvania.

DANIEL H. CHRISTIE, Major, of Granville County; commissioned July 10, 1861; promoted.

E. J. CHRISTIAN, Major, of Montgomery County; commissioned May 10, 1862; killed May 31, 1862, at Seven Pines.

CHARLES C. BLACKNALL, Major, of Granville County; commissioned May 10, 1862; promoted from Captain of Company G.

ISAAC JONES YOUNG, Adjutant, of Granville County; commissioned July 10, 1861; wounded July 1, 1862; promoted Captain of Company G and resigned in 1862.

VINES E. TURNER, Adjutant, of Granville County; commissioned May 10, 1862; wounded at Cold Harbor June 27, 1862; promoted to Captain and Assistant Quartermaster June, 1863.

JUNIUS FRENCH, Adjutant, of Yadkin County; commissioned June, 1863; killed July 1, 1863, at Gettysburg.

CHARLES P. POWELL, Adjutant, of Richmond County; commissioned July, 1863; killed May 9, 1864 at Spottsylvania Court House.

LAWRENCE EVERETT, Adjutant, of Richmond County; commissioned May 12, 1864.

EDWIN G. CHEATHAM, Assistant Quartermaster, of Granville County; commissioned July 10, 1861, resigned February, 1862.

W. I. EVERETT, Assistant Quartermaster, of Richmond County; commissioned in 1862; resigned.

VINES E. TURNER, Assistant Quartermaster, of Granville County; commissioned June, 1863.

JAMES F. JOHNSTON, Assistant Commissary, of Lincoln County.

THEOPHILUS MOORE, Chaplain, of Person County; later Rev. Mr. Berry.

Robert I. Hicks, Surgeon, of Granville County; T. C. Caldwell, of Mecklenburg County, Assistant Surgeon; later Dr. Jordan, of Caswell County, killed at South Mountain.

William F. Gill, Sergeant-Major, of Granville County; killed July 1, 1862 at Malvern Hill.

Charles P. Powell, of Richmond County; appointed May 10, 1862; promoted to Adjutant May 9, 1864.

On 20 May, the day on which North Carolina seceded from the Union, the Confederate Capital had been removed from Montgomery to Richmond. It was now plain that the Old Dominion would be the theatre of the war. Thither the regiment was soon ordered, to return as an organized body no more, with one brief exception, till the great drama of blood and ruin had to the last scene been acted.

On Wednesday, 17 July, Colonel Hoke, with seven companies of the regiment, left the "Camp of Instruction" at Garysburg, N. C., in freight cars for Richmond, Va. Companies C, D and H, were for the time being necessarily left behind on account of the prevalence of measles among the men. Of this malady and in the person of John H. Harmer, Company D, the regiment lost its second man, the first man being Wm. Lowman, of Company A, who died while in camp at Raleigh.

Four nights were spent in camp at "Rocketts" in the suburbs of Richmond. It was either here, or just before leaving Garysburg, that arms and ammunition were first issued to us. The arms consisted of smooth bore percussion muskets, with bayonets; the ammunition of paper cartridges, containing ball and powder. A little later in the war we were armed with rifles captured from the enemy.

MANASSAS.

Early on 21 July, a bright, hot Sunday, our seven companies entrained hurriedly in "box" cars for Manassas Junction. Enthusiasm was at flood tide in that period of boundless hope. Cheers greeted us on every side as we steamed forward and at the stations we were fed and feted. All knew that a battle was impending and later, by means of the telegraph line along the railroad, that it was being fought.

We were eager to go forward; more eager, perhaps, than
we were to reach later fields when experience had unmasked
the true, grim visage of war. But many delays occurred.
The running of the train was so erratic that the engineer was
suspected of treason, though apparently without evident
cause. The soldiers who crowded the tops of the cars in their
eagerness to assist, put on brakes too hard. This caused one
of the car trucks to take fire from friction, or come very near
it. As some of the cars carried, or were believed to carry
powder, the men stopped the train by means of the brakes and
cut the endangered car loose till it cooled.

But these delays were inconsiderable, compared with the
long stop near the Rappahannock bridge, above Gordonsville.
We had started full early and could have reached Manassas
by noon, or soon after. The presence of 700 men, fresh on
the field, might have had great weight at more than one juncture of that dubious battle. But we were sidetracked to meet
many trains of wounded, which began to pass us at Louisa
Court House. During Sunday night we pulled into Manassas Junction. Monday was a rainy, chilly, dismal day.
The men had stopped their cheering and horse play when the
cars of bloody-bandaged wounded passed them the day before
at Louisa Court House. The night spent on the hard car
floors seemed a real hardship. The twenty-four hours fast—
we had left Richmond too suddenly to prepare rations—
seemed then to border on the heroic. The Manassas water
reddened by contact with the mud, then knee deep around the
station, drank like blood. The rows of untended wounded
who had lain all night on the field in the rain, some of them
horribly mutilated, grew longer and longer as the ambulances
came and went. The pile of amputated limbs, naked and
whitened by the chilling rains, grew higher and higher out-
side an amputating tent hard-by the roadside. It was probably the most miserable and trying day that the regiment
spent during the war. The time when the Confederate soldier was to become the marching, fighting, fasting machine
that he did, insensible almost to hunger, cold and mental depression, was yet some distance ahead.

We went into camp at Camp Wigfall, one and a quarter

miles from the Junction. The three companies left at Garysburg under Major Christie, broke camp there on 5 August, and after a few days delay in Richmond waiting for transportation, rejoined us here. For several weeks encamped at this place, the regiment suffered exceedingly from the diseases which then, and even now, seem unescapable by the unseasoned soldier. By the surgeon's statement, the sick call at one time numbered 240, 57 of the cases being typhoid fever. The mortality was large.

After spending several weeks here our first march was made to Camp Ellis, five miles distant, where we remained six weeks. Near here, at Sangster's Cross Roads, our first picket duty was performed. A little later at Mason's Hill the whole regiment went on its first picket. 17 September we pitched camp and began a long stay at Union Mills. Near here, on Bull Run, we built log huts and went into winter quarters in December, where we remained with only such changes in position as the exigencies of the situation in outpost and picket duty required. This gave us an opportunity to enjoy the boundless hospitality of the people of this part of Virginia, upon whom the iron hand of the war was soon to fall with such crushing weight.

Meantime the regiment had been brigaded with the Fifth North Carolina "State Troops," Colonel Duncan K. McRae; the Twentieth Georgia, Colonel Smith; the Twenty-fourth Virginia, Colonel Jubal A. Early; and the Thirty-eighth Virginia, Colonel Jubal Earles. Colonel Early being the ranking officer, he was placed in command, and subsequently commissioned Brigadier-General. General Earl Van Dorn commanded the division, General Beauregard the corps, General Joseph E. Johnston the army. The army was then known as the Army of the Potomac—later upon the abandonment of that line of defense, as the Army of Northern Virginia. In the fall and winter of 1861, many changes took place in the Line Officers of the regiment.

The winter was a severe and trying one. After January 1, 1862, snow, hail, sleet or rain fell almost every day. Frequently all fell the same day. War doffed her holiday mask worn during the tramping from camp to camp, and from picket to picket post in the splendid weather of the past Autumn. Such duties now imposed hardships of a serious and often dangerous nature. Not yet hardened to endure all things, as in time they were, the men suffered intensely from exposure. Great was the mortality from pneumonia, typhoid fever and other diseases.

THE RETREAT FROM MANASSAS.

The early spring of 1862 found the Confederate army along Bull Run and north of that stream, less than 50,000 strong. The Federal hosts under McClellan, confronted it over 100,000 strong. Before the opening of Spring rendered military operations feasible on a large scale, General Johnston decided to withdraw from his exposed position to a stronger line south of the Rappahannock. There he would also be in better position to meet and check any advance of the enemy whether direct or circuitous, as subsequently proved.

The beginning of the retrograde movement found the regiment on picket duty at Burke's Station, on the Orange and Alexandria Railroad, and in close proximity to the enemy who were encamped in the neighborhood of Alexandria and Springfield. The old camp on Bull Run was abandoned 8 March. We moved out at daylight, throwing away tents and camp equipage; sum total of the first day's march one and a half miles, progress being checked by confusion of orders. Early was now acting as Major-General, in command of the Fourth division.

Not until sunset of the 9th, did the grand column move again, reaching Manassas Junction that night. The last we saw of the famous stone bridge across Bull Run, it was in flames. Strictly speaking, stone bridge was a misnomer, all but the abutments being of wood. An immense amount of property was destroyed as the necessity of change of base to the Peninsula was now anticipated. A very carnival, restrained to some extent by the military discipline, reigned that night at the junction. The soldiers got rich with plunder. Depots of supplies and the express office were fired.

Barrels of whiskey were opened at the head and their contents poured in streams on the ground. A rough soldier was observed with six canteens of whiskey around his neck, as if "he wept such waste to see," actually wading in a puddle of the joyful, while in a ditty, tuneless, but gay, he whistled his regrets over "departed spirits."

The next position was south of the Rappahannock. Large numbers of refugees accompanied the army in the retreat. Details of our regiment, as from others, were made to guard and as far as possible, aid them in their wild flight. As the command waded the Rappahannock it witnessed a distressing accident to one of the unfortunates—a widowed lady, half frantic lest she be left behind and taken by the Yankees, missed the ford in driving across the river and was swept down to death by the rapid waters.

For several weeks the army remained in position south of the Rappahannock awaiting a further development of the Federal plans. Then came a long, slow, impeded march along the Orange and Alexandria Railroad. How like a sealed book to the private soldier are the plans of his leaders. How futile our conjectures as to the purpose of our move and the objects to be gained by it. Many yearning hearts—in which the wish was father to the thought—saw in this southern trend only a return to North Carolina.

7 April, we took the cars at Orange Court House and that night, a dark and rainy one, found us in Richmond. After a hastily eaten midnight supper, prepared for us in the market house by the exhaustless hospitality of the good people of Richmond, we were marched to the Yorktown depot. This was the first intimation of our destination. Going by rail sixty miles to West Point, we here took schooners for Yorktown, thirty miles below.

THE PENINSULA CAMPAIGN.

8 April, one month after the beginning of the withdrawal from Manassas the regiment, with other commands, reached Yorktown. Here we got our first experience of the trying duties of life in the trenches, including much toil with pick and shovel. On the 17th, after nine days behind the

breastworks, the boys had their first experience with cannon balls and bomb-shells. The opposing batteries were about three-fourths of a mile apart. The pickets were in rifle pits several hundred yards in advance, and on that day more than one shell exploded in uncomfortable proximity to them. When the first shot was fired direct at the position occupied by the Twenty-third (then the Thirteenth), the writer (H. C. Wall) was on duty in the rifle pits as Sergeant in command. Well is remembered the "sensation" produced by the first shell that fanned the cheeks of ye innocent braves who occupied those rifle pits, and particularly the moving effect wrought upon a certain tongue-tied individual, whose deportment now, as contrasted with previous pretensions, presented a striking consistency with the spirit of the ancient ballad:

"Nought to him possesses greater charms
    Upon a Sunday or a holiday,
Than a snug chat of war and war's alarms,
    While people fight in Turkey, far away."

For, with a precipitate bound, the tongue-tied warrior made tracks for the breastworks exclaiming, in answer to remonstrances and threats of court-martial: "Dam 'fi come here to be hulled out this way when I can't see who's a shootin' at me"—using the terms hulled instead of shelled as synonymous, though he hardly thought of it at the time. At a period a little later in the service, such conduct would have been most severely punished. But it is not remembered that "Dam 'fi" got more than a sharp reprimand and orders for an instant return to his post. If he ever afterwards flinched, we were not informed of it. He was killed at Gettysburg.

As the sharpshooting grew hotter the pickets could be posted and relieved only at night. The opposing pickets fired at everything in sight. For a space the boys on such duty embraced mother earth more intimately than they had before deemed possible. But they gradually learned that shooting and hitting were by no means synonymous terms. At length before the evacuation some of them, at least, preferred a prone position out on the open to the pits half filled with water by the almost incessant rains. The trenches themselves filled with water and could not be drained. Yet the artillery and rifle fire of the enemy held the men close down in them. No fire could be kindled day or night without its becoming the focus of heavy shell fire and it was therefore strictly forbidden. The only food was flour and salt meat and these in diminishing quantities. Food was cooked by details in the rear and brought forward to us. Men sickened by thousands. Soldiers actually died in the mud and water of the trenches before they could be taken to the hospital. And as many of the cases of illness were measles, this exposure meant death. Thus unavoidably died a dog's death many a gallant fellow, who, if spared, would have upheld with his life the Confederate standard, through thick and thin, and to the bitter end. It is not death amid the rapture of the fray that makes war most horrible, but the passing within the dark door of such men under such circumstances. Yet the term of service at Yorktown was not all irksome, nor was it unmarked by occasional diversions from the tread-mill routine of duty. About the quaint old town were many points of interest that awakened patriotic contemplation. The marble slab half a mile from town, marking the spot where eighty years before Cornwallis had surrendered to Washington, was a favorite place of visitation.

Standing there on consecrated ground many a fond prayer was breathed that this self-same spot which witnessed the achievement of American Independence might also see the accomplishment of Southern Independence.

The comparatively insignificant Confederate force at Yorktown had now held McClellan's vast army at bay for weeks, while troops were being concentrated higher up for the defense of the Southern Capital. The Confederate position exposed as it was to turning movements by the Federal fleet on both flanks was clearly untenable. The sole object of Southern strategy, after General Johnston made personal inspection of the surroundings, was simply to check the invasion till the above concentration was completed.

This having been accomplished and holding the enemy in check longer, being possible only by a pitched battle, which it was not desired to fight, the Southern forces were quietly withdrawn 4 May. A deed which, in the heroic days to come, would have passed unnoticed, impressed the unseasoned soldiers, and is yet remembered by many. On the day of the evacuation, part of the Twenty-third were in the rifle pits, which were that day subjected to a fire of unusual keenness. When the officers in the trenches knew that the retreat would begin that night, there was some apprehension that the men in the rifle pits should be captured unless given exact orders what to do. For this purpose Captain C. C. Blacknall, Company G, left the shelter of the trenches under a ceaseless fire at 400 yards, made the circle of the pits, gave the men their orders and returned unharmed. The detail for picket duty from our regiment was the last to leave the works, being relieved by the cavalry at midnight. We marched all night. At dawn when six miles out we heard the furious cannonading of McClellan's assault on our empty intrenchments.

BATTLE OF WILLIAMSBURG.

The retreat, which was much impeded by the slow movement of the wagon trains over the miry roads, was tardy and tedious in the extreme. The ancient town of Williamsburg, in Colonial days the Capital of the Old Dominion, stands only twelve miles from Yorktown. The afternoon of 5 May, a rainy day in the midst of the proverbial cold, wet spell in May, found us only a mile or so above Williamsburg, waiting to see if our aid would be necessary in the expected battle.

From this point Early's Brigade—now composed of the Fifth and Twenty-third (then Thirteenth) North Carolina, the Twenty-fourth Virginia and the Second Florida Battalion—were ordered back to aid Longstreet in resisting the inconveniently eager pursuit of the enemy, for part of the trains were stalled in the deep mud where they stopped the night before, and must be protected or abandoned. The battle was fought on almost the same ground on which the Americans and British contended in 1781. We passed at double quick through the muddy streets of the historic town, pained at the shrieks of women and children who were terrified at the bloody drama then going on in their full view. A short pause to deposit in the campus of classic William and Mary College all knapsacks, extra plunder, etc., none of which we ever saw again—and we are out upon our first battle field.

The design was a charge by Early's Brigade against a strong position manned by Hancock's Brigade on the enemy's right. When drawn up in line for the forward movement, General Early rode the length of the brigade using, in that fine-toned voice of his, something like the words: "Boys, you must do your duty." The line advanced a hundred yards or more through a wheat field wet with the cold rain which had fallen that day, but which had now ceased. Then our regiment was confronted by a forest of trees and thick undergrowth. The line at once became irregular and more or less jumbled by the reason of the natural obstacle to its progress. These woods also shut out the view and caused the line of the regiment's advance to be slightly deflected to the left, by which it lost touch with the Fifth, on our right. At this moment General D. H. Hill appeared, mounted, in our front, and said sharply to the men, now endeavoring to regain their alignment, and each one commanding his fellow, "hush your infernal noise."

In one instant more the right wing of the brigade, having greatly the advantage of the ground in marching, came first in view of the enemy's battery, and charging forward in the open, outstripped the movement of the Twenty-third, impeded by the woods, received a withering fire and was hurled back by a fury of shot and shell irresistible by mortal force. The Fifth North Carolina made a gallant, but fruitless charge, losing many gallant lives, and our regiment was not on hand to support it at the critical moment. That moment was of the briefest possible span—like a sea wave against the sea wall, the charge bounded back almost instantly.

Colonel D. K. McRae, of the Fifth North Carolina, alleged that the Twenty-third (then the Thirteenth) was inexcusably derelict in duty and that Colonel Hoke halted the regiment without orders. Colonel Hoke, on the contrary, maintained that General Early gave the order to halt, which assertion was never denied by General Early. Whether the order to "halt" was given us before or after the batteries opened on the assaulting line, would be hard to tell, for this halt of the regiment appeared to be about the same moment that a portion of the assaulting forces were rushing pell-mell back from the attack. It was all the work of a few minutes and the brigade, chagrined by defeat, and mourning the loss of many gallant spirits, fell back in good order. The enemy seemed content to hold his own, without much further effort to advance his line as night came on. Only four or five men in our regiment were wounded, and all but one of them by random bullets. Captain C. C. Blacknall, Company G, in eagerly leading his company forward through the woods, got some distance in advance, where he came suddenly upon two Federals lying down in the brush. Receiving untouched the fire of one at three paces, he sprang forward with his sword and made them prisoners. The ball that missed the Captain struck James A. Gill, of Company G. This was the first wound of the war received by a member of the Twenty-third. Mr. Gill recovered from his wound and still, at the end of thirty-eight years, survives.

General Joseph E. Johnston, in conversation with me (H. C. Wall) several years after the war, placed the responsibility of the charge upon General D. H. Hill. He said that he did not order it to be made and permitted it only after repeated requests from General Hill. Much was said at the time, and afterwards, of the part our regiment took in the battle of Williamsburg. Blunders there may have been, blunders unavoidable by a command manoevering under such circumstances and amid the exigencies of real warfare for the first time; but the writer of these lines (V. E. Turner) was present as one of its Line Officers, and had every opportunity to be fully conversant with the spirit that animated the regiment. He was conversant with it, and he knows that officers and men were as willing, and even as eager to do their duty as any command in the Southern army. The well known tendency of a man or body of men, endeavoring to go straight forward, but unguided by any distinct objective ahead, as we were in these woods, to bear unconsciously to the left, had possibly had its effect on the deflection in our advance and our separation from the regiment on our right.

Wet as rain can make us, with knapsacks and every shred of extra clothing gone, we marched back to the brow of the hill, where we first formed in line of battle. Here amid mud and rain we were held in line of battle till 3 a. m. As there was momentary expectation of attack, not a spark of fire was allowed. Then twelve miles were tramped, or rather stumbled, through darkness, mud and slush, before halt was made for rest or sleep. The tenacious mire was often knee deep. Shoes were pulled from our feet by it and lost. Pantaloons became so caked and weighted with mud that many, in sheer desperation and utter inability in their exhaustion, to carry an extra ounce, cut off and threw away all below the knees. All that night we had no food, nor the next day, though lunging desperately forward over virtually impassable roads. The following day, the 7th, found us still marching and fasting, or rather, famishing. Blessed indeed were the squad or two that found and shot a razor-back hog. But we were the rear guard and even razor backs had become scarce and wary after being hunted by the 30,000 hungry mouths that had preceded us. One of our Captains who was lucky enough to get an ear of corn a day, always spoke of it as the parched corn march.

Many of the troops "caved" in from sheer exhaustion and starvation. The case of Sergeant Malcolm Nicholson, Company D., which occurred a little later in the retreat, will illustrate our sufferings as well as the grim resolve of the men to keep up with the colors up to the point of absolute physical collapse. This stripling refused to succumb or fall out till at a halt one night he toppled over. His comrades tucked him away in an old wagon body lying near. When the order to "fall in" came, and they went to arouse him, they found that death had given him his discharge and that the weary marching of the boy sergeant was over forever.

On the evening of 9 May, the Chickahominy was reached, the wagons overtaken and the worst hardship of the march, whose sufferings remained ever vivid to the men who clung to the fortunes of the Confederacy to the bitter end, was over.

THE REORGANIZATION.

While camped on the banks of the Chickahominy at Barrett's Ferry, the regiment was re-organized. This was hastened in order to take advantage of a provision in the Confederate Conscript Act, passed 16 April, 1862. This provision allowed troops whose term of enlistment had not expired, to re-organize with all the privileges, as to election of officers, which they had before the act was passed, provided the reorganization was effected within forty days from the passing of the act. With that period lapsed the Confederate soldier's right to choose his own officers, all commissioned officers being thereafter appointed by the President of the Confederacy.

Thus a re-organization of most of the Volunteer North Carolina regiments in that army, a perilous thing in face of a vastly superior enemy, took place about this time, an event unparalleled in the annals of history. A large proportion of officers failing of re-election, their places were filled with men raised from the ranks, or from subordinate positions. Nearly, or quite all the commands, had in their ranks plenty of men competent to serve as commissioned officers. But many thus elevated were not qualified by sufficient experience for command, and the presence of so many inexperienced officers told against the South a month later in the prolonged death grapple with the enemy in the Chickahominy swamps, known as the Seven Days' Fighting. That under such circumstances victory should have crowned Southern effort, attest the dauntless valor of Southern troops.

Our boys, prompted more perhaps by the desire for change, a strong factor in all lives and strongest of all in the monotonous life of a soldier, elected as a rule, new Line Officers.

The following change was made in Field Officers: Daniel H. Christie was elected Colonel in place of John F. Hoke; Robert D. Johnston, formerly Captain of Company K, Lieutenant-Colonel; Ed. J. Christian, former First Lieutenant of Company C, Major; Vines E. Turner, former Second Lieutenant in Company G, Adjutant. That night the officers who had failed of re-election bade us farewell, took leave for Richmond and later sought, most of them, other positions in which to serve their struggling country. Our regiment formerly the Thirteenth North Carolina Volunteers, was thereafter known as the Twenty-third North Carolina Troops.

In pursuance of our plan to briefly outline the careers of the Field Officers of the regiment, we give the following sketch of John F. Hoke, the retiring Colonel.

COLONEL JOHN F. HOKE.

Colonel Hoke was born in Lincoln County, N. C., 8 May, 1820. He was a graduate of the University of North Carolina, and a lawyer by profession. He served with credit as First Lieutenant in Captain W. J. Clarke's company in the Mexican war, taking part in the campaign which resulted in the capture of the City of Mexico. Subsequently he served several terms in the Legislature. At the outbreak of the War for Southern Independence, he was appointed Adjutant-General of North Carolina, serving till the ten regiments of "State Troops" and thirteen regiments of "Volunteers" were organized and equipped. In July, 1861, he was elected Colonel of the Thirteenth (later Twenty-third) North Carolina Volunteers, and commanded the regiment until its reorganization, 10 May, 1862. Failing of re-election, he returned to North Carolina and in 1864 became Colonel of the Seventy-fourth Regiment, Second Senior Reserves). The close of the war found him guarding prisoners at Salisbury. He died in November, 1888. Colonel Hoke was an upright, honorable and cultivated gentleman. Great kindness and consideration characterized his bearing towards the subordinate officers of his regiment.

LIEUTENANT COLONEL JOHN W. LEAK.

John W. Leak was born in Richmond County, N. C., 16 March, 1816. His grandfather, Walter Leak, Sr., served throughout the Revolutionary War as a private in the American army, and died in the town of Rockingham, in 1844, at an advanced age.

He graduated at Randolph-Macon College about 1837.

In July, 1861, he was elected Lieutenant-Colonel of our regiment. This office he filled till the re-organization of the regiment in May, 1862, when, as was the case with many of the officers, he failed of re-election. Being then well advanced into middle age, he retired to private life and became prominent in the cotton mill interests at Rockingham. He died in May, 1874.

THE BATTLE OF SEVEN PINES.

The retreat from the peninsula and up the south banks of the Chickahominy, brought us within sight of Richmond on Sunday, 18 May. We pitched camp in a dense undergrowth of woods, one and a half miles from the city, on the eastern side. Soon the invading Federal hosts drew nearer. Day by day portents of a desperate strife to come, accumulate. Picket firing grows heavier and more persistent, and the shriek and roar of bursting shells seemed to have become part of the natural order of things.

The strategy of the battle of Seven Pines, or Fair Oaks, as it is sometimes called, was exceedingly simple.

McClellan had thrown Keyes' Corps, composed of Casey's and Couch's divisions, and Heintzelman's composed of Hooker's and Kearney's divisions, to the southern bank of the Chickahominy, and Casey had advanced to Seven Pines and fortified. Couch's line was about a mile and a quarter in the rear of Casey's. Hooker and Kearney were in rear of Couch. On Friday night, 30 May, a violent thunder and rain storm had greatly swollen the streams, and Johnston seized upon this opportunity to deal with his vastly superior foe in detail. He hoped to crush these isolated divisions before more troops could be thrown across the swollen Chickahominy to reinforce them. D. H. Hill's division, supported by Longstreet's, was to attack in front; Huger's division was to attack the enemy's left flank, and Smith's his right.

The Twenty-third took an important and most gallant part, both in the battle of Seven Pines and in the reconnoissance on the Williamsburg road the day before, which disclosed the situation of the enemy and led to the Confederate attack. In this sortie down the Williamsburg road 30 May, several men were wounded and Captain Ambrose Scarborough, of Company C, in command of the four companies reconnoitering, was killed. In the person of this gallant officer the regiment lost its first man from a hostile bullet. Captain Frank Bennett commanded the advance line of sharpshooters, who really developed the enemy's strength, was severely wounded, being disabled for months.

In the attack at Seven Pines, made in the afternoon of Saturday, 31 May, 1862, the Twenty-third belonged to Garland's Brigade. This with three other brigades, Rodes', G. B. Anderson's and Raines', formed Hill's division, which assaulted the strongly fortified Federal front. Few attacks in war were ever made under circumstances more unfavorable to the assaulting force. A swamp, in some places waist deep in water and thick with undergrowth and tangled vine, had to be crossed, and a skillfully made abatis confronted and struggled through before the heavily manned hostile works beyond could be reached. Through them all swept the regiment in line, with its comrade commands, under a fire of musketry and artillery as hot as mortal men ever breasted with success. Many a gallant fellow was stricken down dead or wounded. Some rendered helpless by wounds, not necessarily fatal, sank and were drowned in the deep waters of the swamp.

Finer tribute to fighting men was never paid than that by a Northern writer who saw the battle from the point of view which we assailed—there being no hotter section of that fire-swept line than which fate assigned to the Twenty-third. This writer says: "Our shot tore their ranks wide open, and shattered them in a manner frightful to behold, but they closed up and came on as steadily as English Veterans. When they got within four hundred yards we closed our case shot and opened on them with canister. Such destruction I never witnessed. At each discharge great gaps were made in their ranks. * * But they at once closed and came steadily on, never halting, never wavering, right through the woods (swamp), over the fence, through the field, right up to our guns, and sweeping everything before them, captured our artillery and cut our whole division to pieces."

Huger's turning movement far to our right had been stopped by impassable streams. Smith's attack far to our left, where General Johnston commanded in person, had been beaten off, and the Commander-in-Chief severely wounded.

But in our front the victory was complete. After two hours, ending in the brilliant charge described above, Casey's works were carried and his routed line driven back on Couch's. Then the division reinforced by only one, R. H. Anderson's, smashed Couch, though reinforced by Kearney, and drove all back on their third line two miles in rear of the first line. Twelve pieces of artillery and 6,000 stands of small arms, were taken. Darkness put an end to the battle.

But a heavy blood equivalent was paid for the victory. Owing to much sickness the regiment, according to the statement of Captain A. T. Cole, was able to go into this action only about 225 strong. Moore's Roster, which in countless instances, and probably in this, is incomplete, shows that twenty-four privates and non-commissioned officers were killed, and ninety-five wounded, sixteen of them mortally. As will be seen, this was an exceeding large proportion of the number engaged.

There were also many casualties among the commissioned officers. None of the Field Officers escaped injury. Colonel Christie was wounded. Lieutenant-Colonel R. D. Johnston was wounded in the arm, face and neck, had his horse killed under him and was shot down within fifty feet of the hostile works. Captain C. C. Blacknall, Company G, who, unable to walk, owing to a sprained ankle, had gone into action mounted, was grazed by seven balls, and received a painful bruise near the spine from a fragment of shell. He also received painful injuries from his horse, which was killed and fell on him. Captain William Johnston, Company K, and Lieutenant E. A. McDonald, Company D, were also wounded. Lieutenants J. D. Knott and A. M. Luria, of Company I, were killed. Luria was a gallant young fellow. It was at Seawell's Point that he did a heroic act, which, had he been a British soldier, would have brought him the Victoria Cross and caused the world to ring with his name. While there early in 1861, either as a visitor or as a member of Colquitt's command, before he joined the Twenty-third, a shell from the Federal gunboats dropped among the Confederates. With rare presence of mind and devotion, he seized the shell and threw it over the works before it could explode. At our reorganization he refused promotion, saying that he wished nothing unless won on the battle field.   Major E. J. Christian was mortally wounded, dying a few days later.

MAJOR EDMUND J. CHRISTIAN.

Major Edmund J. Christian was born in Montgomery County, N. C., in 1834. His uncle, Samuel H. Christian, was elected to the Confederate Congress, but died before taking his seat.   While a boy, his father died, leaving his mother and her other children largely dependent on him, which duty he successfully performed.   Major Christian was a farmer by vocation. He was a man of magnificent physique and had no bad habits.   On the outbreak of war he enlisted as a private, but was elected Lieutenant, in the Montgomery Vounteers No. 1, which became Company C on the organization of the regiment.   Upon the reorganization, 10 May, 1862, he was elected Major, to fall in battle just three weeks later.   At Seven Pines he had received two wounds, either of which would have justified his retirement from the field. But he pluckily went forward at the head of his men till stricken down with the third and mortal wound.   He was conveyed to a private house in Richmond, tenderly nursed for the two or three days he had to live, and was laid to rest in the Confederate Capital which he had died to defend.   Lieutenant W. P. Gill, of Company G, was also wounded.

Captain C. C. Blacknall, Company G, was promoted to Major on the death of Major Christian.

The courage and dash of the men and officers in this bloody onslaught, has never been surpassed.   When in the impetuosity of the onset through the vine-tangled swamp, the three right companies became temporarily separated from the regiment.   Lieutenant-Colonel Johnston led them gallantly forward with the Fourth Regiment.   Splendidly did the whole command show its alacrity to meet and close with the foe, no matter what the obstacles, so that they knew where he was and there was no confusion of orders as in the woods at Williamsburg.   The conduct of private Wm. C. Cole, brother of Captain A. T. Cole, at Seven Pines, is a good illustration of the high resolve of the men to do their full duty. This youth, a mere stripling and in poor health from the hardships of the campaign, found in the thick of the fight, that the channel of the tube was obstructed, and that his musket would not fire, sat down under a hot fire, removed the tube with his wrench, screwed home a new one, caught up with the line at a few bounds and continued to load and fire as long as a Yankee was in sight.

After Seven Pines, the regiment went into camp near Richmond and passed several weeks in drilling. Here on Tuesday, 17 June, it was re-brigaded, being now placed in brigade with the Fifth, Twelfth, Thirteenth and Twentieth, all North Carolina regiments. Samuel Garland, Jr., of Lynchburg, Va., remained in command as Brigadier. Soon after the wounding of General Joseph E. Johnston at Seven Pines, General R. E. Lee became Commander-in-Chief of the army.

THE SEVEN DAYS' FIGHTING.

As the month of June, 1862, wore away, McClellan's plans developed. The Confederate Capital was to be taken by regular approaches. The 26 June found his splendidly organized and equipped army of at least 105,000 effectives, strongly intrenched on a line straddling the Chickahominy and extending from White Oak Swamp, twelve miles southeast of Richmond, to Mechanicsville, six miles northeast. The line, especially that part north of the Chickahominy, ran along positions of great natural strength, rugged bluffs protected largely by streams or swamps on the side next to the Confederates.

The southern strategy of this protracted death grapple, so well described by its name, the Seven Days Fighting, was masterly—as brilliant as history records. The valor and staying powers evinced by the Southern soldiers in that prolonged combat is scarcely matched in the annals of time. But for an apparently inherent defect in the Southern mind—its inability to master, or its universal contempt for, the practical details of things, the invading hosts would in all likelihood have met its doom in the Chickahominy swamps. Had Southern practicalness been at all commensurate with Southern generalship and Southern courage, it is hard to see how McClellan's army could have escaped ruin, if not total destruction. This unpracticalness manifested itself here in the failure to prepare accurate topographical maps of a region which the trend of events had, for months, pointed out as the most probable scene of conflict.

The position of the Federal army was, on the whole, naturally very strong and made as much stronger as engineering skill could make it. But owing to the isolating effect of the many streams and swamps, difficult of passage, it gave the opportunity of the war to the qualities in which the Southern army excelled—prowess and military genius. In this instance these qualities were largely negatived by the fact that the Confederate leaders fought and manouvered over a region of whose exact topography they knew scarcely more than of the craters in the moon. The result of this ignorance of natural obstacles, and of the roads that turn them, was that thousands of gallant men, the very flower of the Southern army, were needlessly and heedlessly sacrificed, and that a half victory cost double the price for what a whole victory could have been obtained.

Lee's plans were that Jackson, then in the Shenandoah Valley, by a rapid and secret march, should strike the right flank of this twenty-mile line, while he smote its right front. Then beginning at the end, 55,000 of his 80,000 men, were to be thrown impetuously against the Federal line, flanking it as far as practicable, and rolling it back upon itself, compass its destruction if possible.

After Seven Pines the Twenty-third was assigned a position near the left wing of the army. Our tents were pitched on the banks of a small stream about 600 yards in the rear of the works. As an advance of the enemy was hourly expected, the orders were that upon the sound of a bugle at brigade headquarters, the regiment must be formed in five minutes with three days' rations, canteens filled and forty rounds of ammunition per man, ready to march rapidly to its place in line. This rendered it necessary for the men to sleep with their cartridge belts on and haversacks and canteens by their sides. Mounted officers had to keep their horses saddled. No one was allowed to be absent from the command for a moment. Many such alarms were given by day and by night. Two weeks of this rigid discipline made the order to advance a genuine relief.

The fighting began in earnest on Thursday, 26 June, a fine cloudless day. On the afternoon of that day A. P. Hill moved to the east and without waiting for Jackson's appearance on the Federal flank, as had been agreed, assaulted in front the impregnable lines on Beaver Dam Creek, a small stream running north and south, and emptying into the Chickahominy. The result was that he was beaten off with the loss of over 3,000 men, a loss nearly ten times as great as he inflicted on the enemy. This is often called the battle of Mechanicsville from a very small village at the cross roads a mile west of the stream. [This premature assault and consequent disastrous and useless loss of life General A. P. Hill afterward repeated at Gettysburg and at Bristoe Station.—Ed.]

The Twenty-third, which belonged to D. H. Hill's division, was not actively engaged on the 26th. About 11 a. m. of that day, we left our position in line and marched to the left, striking the Mechanicsville road as we filed down the hill towards a little stream. To the left of our line of march could be seen a group of high Confederate officers, including President Davis, Generals Lee, Longstreet, D. H. Hill, Garland and others. Their earnest consultation and the distant firing made us feel that a momentous period in the struggle was now at hand. We were marched up and took position opposite the hills beyond the stream, and were for a while under a spirited cannonade. Adjutant Turner's horse was killed, falling on him, but not inflicting injury enough to keep him out of the battle of the next day. Several other casualties were also sustained by the regiment.

We slept that night on our arms. Early the next morning while Captain I. J. Young was getting his company in line for the work before us, one of his men complained that he was not well, and wanted to report on the sick list. Captain Young was heard to say: "Yes, damn it; I know you are sick. But it's only the battle field colic. I'll not excuse you." The diagnosis proved correct, the "colic" soon passed and the patient, we believe, did his duty faithfully that day.

Upon the approach of Jackson from the north on their right flank, the enemy withdrew from their strong line on Beaver Dam Creek, to one scarcely less strong on Powhite Creek, another small stream running parallel with Beaver Dam and about four or five miles to the east of it. A. P. Hill, Longstreet and D. H. Hill followed closely.

A little to the east of Powhite Creek was fought the battle called Gaines' Mill, and less commonly the battle of Cold Harbor. But for the fact that it would be confounded with the battle fought there on May, 1864, the latter term is more accurate, for the enemy were brushed back from the line at Powhite Creek on which stands Gaines' Mill with comparatively little fighting. Their stand to the death was made behind a great semi-circle of swamps a mile or more to the east of Powhite Creek, and much nearer New Cold Harbor than Gaines' Mill. On the morning of the 27th, D. H. Hill's division was thrown forward, well to the left along the road running by Bethesda Church, so as to reach Porter's right rear. When, after much delay and perplexity, at 2:30 p. m., we came into collision with the enemy near old Cold Harbor and three miles northwest of New Cold Harbor, our brigade, Garland's, was on the extreme left of the enemy.

It was nearly sun down when the two brigades of Anderson and Garland got permission from D. H. Hill, their division commander, to advance to the charge. The assault was delivered under conditions not unlike those at Seven Pines nearly a month earlier. A swamp densely covered with undergrowth had to be passed under fire before the Federals could be reached. These consisted of United States regulars under Sykes, a hard and persistent fighter.

But nothing could withstand the impetuosity of our onward sweep. Alignment was soon lost in the contraction of the lines necessary in attacking a shorter front than our own. But the Twenty-third, along with the other regiments, pressed forward, tearing their way through brush and briar and vine.

After clearing these bewildering obstructions we emerged into a thin piece of woods with no undergrowth. This brought us in full view of a battery on our left, which opened upon us, as we went forward at the double quick down a little slope. The men became excited and began to fire; but Colonel Christie sent his Adjutant, the writer of this, to stop the firing till they got closer. So down we swept and then up the hill to the enemy's position. Just at this juncture came the critical moment of the day, and possibly of the campaign. Their line began to waver. Officers and men seemed by one accord to grasp the situation. We pressed forward in the charge as a part of an Alabama regiment rushed back upon our line. Its Colonel shouted that he was going back to reform. Captain Young, then in command of the regiment, Colonel Christie having just fallen severely wounded, exclaimed: "Don't go back to reform. We are all needed to carry this line." So the regiment turned and charged with us.

Up the hill we pressed. The enemy now broke and fled in great disorder through a dense swamp in their rear, leaving large numbers of knapsacks behind them. We took sixty or seventy prisoners. It was now dark. We were hungry, worn out and entirely separated from the other regiments of the brigade which had gone in and broken the line to the right and left of us.

We bivouaced in a body of pines, too worn out to stand guard over prisoners, who seemed as tired and worn out as ourselves. The Adjutant counted them and cautioned them not to move during the night. Then lying down around them, we slept soundly. They seemed well contented and showed no disposition to escape while with us.

There has been much dispute as to what troops first broke the enemy's line at Cold Harbor, and thus began the long chain of McClellan's reverses. But Northern writers state that the right wing gave way first. This is where D. H. Hill's assault was delivered. General Hill himself says that Garland's charge made the first break in the hostile line. General Lee officially paid high compliment to the division for its part in this battle.

Our regiment was not actually engaged at Savage Station, Fraser's Farm, or any of the subsequent battles, till Malvern Hill, fought on Wednesday, 1 July. McClellan beaten and harried on every hand, saw that escape would be difficult, probably impossible, unless Lee's pursuit could be checked. For this purpose on Tuesday night, 30 June, and early the next morning, he hurried to Malvern Hill his shattered commands. If the hand of Omnipotence, molding plastic nature at will, had contrived a fastness in which a beaten and dispirited army might take refuge and grow strong in a sense of security, it need do no more than fashion another Malvern Hill. Here with the James river to his back, and his fleet of gunboats on his left flank, he felt that he might meet even Lee's dauntless, though shattered divisions. Here, frowning tier above frowning tier, in implacements made by nature's own hand, his 300 pieces of splendid artillery were concentrated. Hither his still formidable army, now as at the beginning, far outnumbering the Confederates, was drawn back and skilfully massed in time to strengthen, with partial entrenchments, the points that were least strong. A clearing of 500 to 900 yards between the Federal position and the woods and Swamp in their front, gave a full view of their assailants.

Against this inland Gibraltar, the Southern troops were hurled. A simultaneous attack along the whole line would have been desperate. Attacks at intervals, at the different points by different commands without concert of action, were hopeless. Yet such, by an unfortunate concatenation of errors, was the mode of attack. Late on that sultry summer afternoon our division (D. H. Hill's) struggled through an almost impassable swamp and opened the battle with the first direct assault. Our brigade (Garland's) was in the first line, and advanced through the broadest part of the belt of cleared ground, which had been broken by the plow on the side next to the enemy. Though only Whiting's small division was to the left of us, our attack was directed against the Federal centre. Here we fought Couch's men which we had routed at Seven Pines and when here, as there, hard pressed, Kearney came to their aid.

But the task now assigned us was beyond the power of mortal men. From the first step in the open, the fire of that huge volcanic amphitheatre and of the gun boats on the river was focused on us, much as the ribs of a fan meet at the handle. Yet onward we swept; the line, when shattered and hurled back in places, reforming and pushing with grim determination, doggedly forward, breaking in part the first line of the enemy. No field ever more fully tested the fibre of Anglo-Saxon manhood, and on no field has it ever acquitted itself better. Not till they had striven, unaided for more than an hour against McClellan's whole army and 2,000 had fallen, did they yield to the inevitable and were swept backward by the moving wall of lead and iron.

As at Seven Pines, we will let foeman pay tribute to their matchless ardor. A French officer, the Comte de Paris, who was on McClellan's staff, saw it all and said the following:

"Hill advanced alone against the Federal position.   *   * He had therefore before him Morell's right, Couch's division, reinforced by Caldwell's Brigade   *   * and fronting the left of Kearney. As soon as they (Hill's troops) passed beyond the edge of the forest, they were received by a fire from all the batteries at once, some posted on the hills, others ranged midway close to the Federal infantry. The latter joined its musketry fire to the cannonade when Hill's first line had come within range, and threw it back in disorder on its reserves.

While it was reforming, new battalions marched up to the assault in their turn. The remembrance of Cold Harbor doubles the energies of Hill's soldiers. They try to pierce the line, sometimes at one point, sometimes at another, charging Kearney's left first and Couch's right   *   * and afterwards throwing themselves upon the left of Couch's division. But here also after nearly reaching the Federal position, they are repulsed. The conflict is carried on with great fierceness on both sides, and for a moment it seems that the Confederates are at last to penetrate the very centre of their adversaries and of the formidable artillery which was now dealing destruction in their ranks. But Sumner, who commands on the right, detaches Sickles' and Meagher's brigades to Couch's assistance. During this time, Whiting on the left and Huger on the right, suffer Hill's soldiers to become exhausted without supporting them. * * At 7 o'clock, Hill reorganized the debris of his troops in the woods * * his tenacity and the courage of his soldiers had only had the effect of causing him to sustain heavy loss."

Not till far in the night did the terrific volcano of Malvern Hill become extinct. Fearful had been its execution not only on the fighting line, but numbers of the supports far back in the woods to the rear had been struck down. It was one of the few battles in history in which the casualties from artillery fire were as large, probably larger, than those from small arms.

Battered and shattered, but undismayed, the Twenty-third slept that night upon its arms ready for the eventualities of the morrow. But the stir and rumble within the hostile line had been significant. Jackson's drowsy response, when awakened from the slumbers which from sheer exhaustion had mastered him, and asked what must be done should Mc-Clellan attack tomorrow. "He won't be there," had been indeed prophetic words. The morrow broke over Malvern Hill tenanted only by Federal dead and wounded, all of which the enemy had left in their flight. It broke over the "Little Napoleon"—very little he then appeared at Washington, if not to himself—safe under shelter of his gunboats at Harrison's Landing, clamoring for 50,000 fresh troops. McClellan had lost 15,849 men in killed, wounded and captured, besides 52 pieces of artillery, 27,000 stands of small arms and millions of dollars worth of stores. But the Confederates being everywhere the assailants, sustained a still heavier loss, their casualties reaching the enormous aggregate of 19,749.

It is impossible to give with accuracy our regiment's loss during the Seven Days fighting. Moore's Roster, often inaccurate and incomplete, is here unusually so. According to statement of Captain A. T. Cole, Company D, who estimates the casualties of the regiment in proportion to those known to have been sustained by his own company, the Twenty-third began the Seven Days fighting with about 175 men. It sustained the heaviest loss at Malvern Hill. Here about 30 were killed and 75 wounded. These figures, while only approximate, are believed to be near the mark. These losses left the command a mere skeleton, till strengthened by recruits and the return of wounded men who had recovered.

Colonel D. H. Christie and Adjutant V. E. Turner were wounded at Cold Harbor. Captain I. J. Young, who commanded the Twenty-third at Malvern Hill was, in that battle, wounded in the face, and Private C. C. Courtney, Company A, killed in taking him from the field. Here also Captain A. T. Cole, Company D, and Lieutenant Munday, Company K, were wounded, and Lieutenant Wm. F. Gill, of Granville County, killed. The list, though incomplete, covers so far as can now be ascertained, the casualties of the commissioned officers.

LIEUTENANT WM. P. GILL.

Wm. P. Gill was born in Franklin County, N. C., October 1842. While yet a lad fresh from college, he enlisted as a private in the Granville Rifles, afterwards Company G, was appointed Sergeant Major and at the reorganization, elected Second Lieutenant in the company. His duties as Sergeant Major had brought him in frequent contact with the officers of his regiment, and most of the men. His death caused genuine sorrow and regret to every member of the command. He was handsome in person, and his bearing that of a gentleman. His bravery, manliness, his frank, open face alight with the quenchless enthusiasm of a youth, won and held the love and respect of all. For though gentle and polite, he was firm in the discharge of his duties. His abilities were so generally recognized that his promotion must have been rapid had he been spared to his country and the army. He said the morning of Malvern Hill, that he would not survive the battle. So strong was this premonition that when Captain I. J. Young was borne to the rear, wounded, he asked the Captain to take charge of a watch which had just been entrusted to him (Lieutenant Gill) by a dying Federal, for transmission to his mother. And I will add that after the war Captain Young found the mother and delivered the watch. Lieutenant Gill, now in command of the regiment, was instantly killed, being almost cut asunder by a shell, after the attack was over.   He was then only nineteen.   With his fall perished one of the noblest spirits of the command.

It was at Malvern Hill that private Charles P. Powell, of Company D, emulated the fearless deed of Luria at Seawell's Point.   While lying in line under heavy artillery fire, waiting for the order to charge, a shell dropped among us. The men could not leave their places in the line of battle, so they flattened to earth while their unwelcome visitor sputtered away.   An instant later the heroic Powell sprung forward, lifted the shell and deliberately sousing the head in one of the small water pools of the swamp, put out the fuse.   The fuse must by some error have been cut a trifle long, or after so much delay it must have exploded in his hands before it reached the water.   This gallant fellow was wounded a little later in this battle and also at Gettysburg, promoted Adjutant and was killed in the "Bloody Angle" 12 May, 1864—an immortal record, surely.   The wounding of Captain Young left Second Lieutenant Gill in command of the regiment till he was killed.   After his fall the Twenty-third seems to have had no commissioned officer left on the field.

After the battle we spent several weeks of grateful and well needed rest near Richmond.   When Jackson, followed later by the bulk of the army, marched against Pope at Manassas, our division was, with McLaws' left behind to observe the enemy and guard Richmond.   In fact, D. H. Hill's division oftener than any other, was detached on independent service of that kind.

When McClellan's army was withdrawn to reinforce Pope and safeguard the Federal Capital, we were thrown forward by forced marches northward.   We rejoined the Confederate army at Chantilly 2 September, three days after the battle of Second Manassas was over.   The earth was yet encumbered with unburied dead.   The most gruesome of our whole war experience were the many swollen corpses crushed and mangled by the cannon wheels, which in the urgency of that fierce and prolonged combat had passed over them.   Artillery must manœuvre somewhere; the dead lay thick nearly everywhere, and men had been too engrossed wielding the sickle of death to gather in the harvest.

THE FIRST MARYLAND CAMPAIGN.

At Chantilly we were within a few miles of the scene of our picket duty the previous Fall, Winter and Spring. But our pause there was of the briefest. Our brigade formed Lee's vanguard in the invasion of Maryland. Moving rapidly northward Friday, 5 September, we waded the Potomac near Leesburg, at Noland's Ford, lower down, we believe, than the Southern army crossed it before or after in its many passages. With what bounding hearts did we climb the opposite banks of the Potomac, looking eagerly for the support of "Maryland, My Maryland." Cherishing hopes which, alas, like so many other Confederate Hopes, withered on the stem.

Strong indeed must have been the Southern proclivities of Maryland men to see aught of attraction in a service like ours. We were a hungry, jaded, weather-beaten, battle-worn set. In the forced marches to the northward wagon trains had been outstripped, green corn and apples forming for days almost our only food. The fields of "roasting ears," most of them now too hard to be really edible, were bought from the farmers and the men turned in to help themselves. One of General Hill's first acts after crossing the Potomac into Maryland, was to buy a large field of corn and turn in his division. All supplies obtained during this campaign were paid for in Federal currency.

A cartoon in Harper's Weekly represented a Maryland Quaker woman placing a wash stand at her door and imploring the rebels that if they must possess her house, please to use that first; while the rebels mistaking this—to them—strange apparatus for some infernal machine contrived for their destruction, skedaddled forthwith. But the lion, though unkempt and half starved, was a lion still, as the foe discovered when he threw his 90,000 against our ranks thinned by battle, disease and the giving out of foot-sore men, to less than 30,000.

76   **NORTH CAROLINA - 23rd REGIMENT INFANTRY - Roster and History**

****

From the 6th to the 10th of September, we remained in camp near Frederick, Maryland. Here rest, full rations and delightful weather recuperated us fast. It was while in camp at this place that the famous "Lost Order" was dropped by some one at the headquarters of our division commander, General D. H. Hill. General Hill subsequently established the fact that he never saw this duplicate order. The soldier who lost it was never guilty of a more culpable act, nor one fraught with more moment. This order, which was picked up on the 13th by a Federal soldier, wrapped around some Confederate cigars, and at once transmitted to McClellan, revealed not only the dangerous secret that Lee's army was divided, but told in minute detail the present position and future movements of infantry, artillery, cavalry and trains. In the hands of an able and active foe—one alive to the tremendous advantage thus given him and quick and resolute in availing himself of it—this paper must in all probability, have been the death warrant of the Southern Confederacy. For by a strange fatality it revealed the faults of Southern strategy at its faultiest moment, and told where and when to meet and overcome the Confederate commands when their strength was at the lowest ebb.

Fortunately McClellan had few of the qualities of a Jackson, a Forest or even a Hood. He acted upon the information thus obtained, but not with the promptitude and energy that Fate demands when at long intervals she places such opportunity in mortal hands. McClellan's report of the find to Lincoln was not only characteristic, but a fine tribute to the valor of his weakened, scattered and now betrayed antagonists. "I have all the plans of the rebels," he wires, "and will catch them in their own trap, if *my men are equal to the emergency.*" Lee's strokes had been so hard and his strategy, based upon the prowess of his army, so bold, that McClellan informed Halleck that he had "120,000 men to fight."

McClellan thus apprized of the situation, moved forward on the morning of 13 September, to take advantage of it.

One column under Franklin was thrown forward southwesterly towards Crampton Gap of South Mountain. Its objective was to crush Jackson's force, then hammering the

Federal garrison at Harper's Ferry. The bulk of the Federal army was moved westerly against us through Turner's and Fox's Gap, its object being Hagerstown, which the "Lost Order" had disclosed as Lee's point of rendezvous.

We had withdrawn from Frederick 10 September, moving slowly through Turner's Gap of South Mountain towards Boonsboro, on the direct road to Hagerstown. Our division was the rear guard of the army and was encumbered with all the wagon and artillery trains.

THE BATTLE OF BOONSBORO, OR SOUTH MOUNTAIN.

By the afternoon 13 September, we had marched to the west of Boonsboro, and gone into camp near Funkstown. From here we were hurried back east to South Mountain; meeting General Stuart coming down as we marched up. That night we spent on the western slope of the ridge; a chilly bivouac without blankets or any manner of covering from the keen mountain air.

Early in the morning of Sunday, 14 September, General D. H. Hill came in person and posted Colquitt's brigade in Turner's Gap and our brigade (Garland's) in Fox's Gap, a mile to the south of Turner's. These two Gaps, which are virtually one, are traversed by many roads. If McClellan's advance was to be checked till Jackson could take Harper's Ferry and join Lee, all these roads must be held by this handful of men against McClellan to the last extremity. This necessitated the scattering of the regiments of the brigade and resulted almost in the destruction of some of them, but the pass was held and the precious time necessary for Lee to concentrate, gained.

Garland's brigade of five regiments numbered less than 1,000 men. Our regiment had been severely cut to pieces at Seven Pines and Malvern Hill, and not yet having been recruited by conscripts to the same extent as some of the others, was much smaller in proportion than the brigade. Our position was in the centre of the brigade and along the crest of the ridge behind an old stone fence, so common in that region.

The fence had been more or less dismantled by time and was in places very low.

To our right was the Fifth and then the Twelfth; to our left the Thirteenth and then the Twentieth. An interval of fully 250 yards separated the Thirteenth from the Twenty-third, and one probably as great severed it from the Twentieth.

Against Garland's 1,000 Cox led 3,000 of Reno's Corps. The action begun at 9 a. m. From our elevated position we had a full view of the movement in our front. Below us in plain view, went forward through the woods the skirmish line of the brigade. Near them and slowly drawing nearer and nearer, came a dark-blue line. Yet they apparently did not see each other. Not till the lines seemed within a few yards of each other was the calm, radiant Sabbath morning broken by the crack of rifles. The battle was on.

Our skirmish line was soon forced backward by weight of numbers. General Garland seeing this, ordered Colonel McRae to take his regiment, the Fifth, and the Twelfth regiment and support the skirmish line. This he attempted to do, but the main line of the enemy coming up at this juncture, forced our skirmish line back in disorder and developed so much strength that McRae not being able to prevent the advance, fell back to his position on our right.

The Federals now pressed forward, striking first the Thirteenth and Twentieth on our left. Here General Garland fell. But as General Hill says, the main attack was against the Twenty-third behind the stone wall (tumbled down stone fence). A little later, but while still fiercely contending on the left, assault after assault was made against our front. These we beat off, inflicting heavy loss on the assailants. At length Colonel Christie seeing that a still stronger force which was advancing against him could, while engaging his front, envelop his left, sent his Adjutant, V. E. Turner (the writer of this) to apprize General Garland of the situation. Finding that Garland had fallen, the Adjutant, making his way towards the rear of the Thirteenth and Twentieth, delivered the message to Colonel McRae, then in command of the brigade. Colonel McRae having no horse or Staff (General

NORTH CAROLINA - 23rd REGIMENT INFANTRY - Roster and History   79

****

TWENTY-THIRD REGIMENT.

1. Geo. Burns Bullock, Captain, Co. I.
2. N. A. Gregory, 1st Lieut., Co. I.
3. Richard V. Minor, 1st Lieut., Co. E.
4. W. P. Gill, 2d Lieut., Co. G.
5. William H. Harris, Private, Co. I.
6. John T. Sanford, Private, Co. I.
7. Nicholas T. Green, Private, Co. E.
8. John H. Breedlove, Private, Co. G.
9. James R. Hart, Sergeant, Co. I.

Garland's Staff having gone off with his body) had no means of immediate communication with General Hill, and was unable to fill the gap and to avert the disaster apprehended by Colonel Christie.

The returning Adjutant after almost running into the hostile lines, reached the position of the Twenty-third just as it was abandoned. Colonel Christie, with his short, weak line, hopelessly enveloped and enfiladed, and seeing capture sure if he remained longer, had ordered the regiment to withdraw. This withdrawal, as it had to be precipitate in the extreme, was effected in great disorder down the steep and bewildering mountain side. Company E and a few other men on the left, the side on which the flank attack came, either did not hear the order to withdraw, or being already enveloped, were mostly captured. It was here and by this gallant Company that bayonets and clubbed muskets were so freely used in the vain struggle to repel outnumbering foes. The regiment had been too roughly handled to be taken into action again that day.

The whole brigade was likewise driven back, though the Thirteenth on the left, managed by a change of front, to maintain itself till reinforced by Anderson's brigade. The exact loss of the Twenty-third cannot now be ascertained, but it was heavy in killed and wounded and of the 200 prisoners captured from the brigade it lost its share. It also inflicted heavy loss upon the enemy before the stone fence, its post of vantage, was enfiladed and rendered useless. General Jesse L. Reno, commanding the corps assailing us, and who had been prominent in the capture of Roanoke Island, Kinston, and other places in North Carolina, was killed at long range by Charles W. Bennett, of Granville County, Orderly Sergeant of Company E. Sergeant Bennett was severely wounded at Sharpsburg. Among our wounded was also Captain G. T. Baskerville, of Company I. General Garland was killed early in the action. In making his way to the firing line, he passed through an open space to the rear of the gap, between the Twenty-third and the Thirteenth. He had been told that the Federal sharpshooters commanded this space, but could not believe that they had yet advanced far enough to reach the crest and dominate the place. Venturing through the opening, he at once became their target and was shot down.

The arrival of reinforcements late in the day enabled Hill, by desperate fighting, to hold Fox's and Turner's Gap till dark, as Crampton Gap, to the south, had been held. Under cover of night all three gaps were evacuated and the Confedate forces concentrated on Sharpsburg, whither Jackson hastened on the fall of Harper's Ferry.

DR. JOURDAN, ASSISTANT SURGEON.

When the enemy at last succeeded in getting in on our left flank and cutting us off from the other regiments of the brigade, Dr. Jourdan was so near the firing line that he was not recognized as a "non-combatant," and was deliberately shot down. He was a native of Roxboro, Caswell County, N. C.; was most highly esteemed as a gentleman and an efficient officer, always kindly and considerate of the sick and wounded.

On the march, when the ambulance was filled with the sick, he often gave up his horse to disabled men and marched on foot himself. The whole regiment were greatly devoted to him.

THE BATTLE OF SHARPSBURG, OR ANTIETAM.

Jackson captured Harper's Ferry 15 September, and by forced marches joined Lee, with most of his forces, at Sharpsburg on the 16th. McClellan advanced and threw part of his command over the Antietam Creek that night. The battle joined at daylight of the 17th. And in that bloody Wednesday was crowded more desperate fighting and more carnage than the New World had ever seen in one day. Retreating along the Boonsboro road, we reached the field early on the morning of the 15th, with the enemy close behind us as we crossed the bridge over Antietam Creek. We at once took position along the ridge and in an open field.

The Twenty-third regiment was here able to muster but few men, many being barefoot and absolutely unable to keep up iu the forced marches over rough and stony roads. The brigade which since Garland's fall, had been under the command of Colonel McRae, of the Fifth, went into action with Colquitt's brigade in the Confederate center, and were advancing in perfect steadiness under a heavy artillery fire from the opposite hills, till the unaccountable "run back" occurred. This happened as follows: The Federals advanced against us in dense lines through a corn field, which concealed the uniforms, though their flags and mounted officers could be seen plainly above the corn tassels. As the blue line became more distinct, approaching the edge of the corn field, which brought it in our range, we commenced to fire and effectively held it in check. But some of Early's men, who had come from the corn field, begged us not to fire, saying that their men were in our front. Some one in a regiment to the right of us also shouted: "Cease firing. You are shooting your own men." Hands were also seen waving the line back. This confused the men. The artillery fire grew constantly hotter. Several of the regiments, nearly exterminated at Williamsburg, Seven Pines and Malvern Hill, had been recruited with raw men, largely ignorant of discipline and of the machine-like duties of a soldier.

At this the regiments on our right began to fall back, straggling through the woods in our rear. But we could plainly see that we were not firing on our friends, and in our front the enemy was firmly held in check, till we found that they were moving on our flank unopposed. This compelled us to retire, which was done in good order, considering the circumstances. The greater part of our regiment stopped in a sunken road (the famous Bloody Lane) and joining the main line there, fought the remainder of the day. General Hill says distinctly that the Twenty-third was kept intact and moved to the sunken road.

The brigade was gotten together that night and early the next morning. The 18th was spent in line of battle ready for the attack which did not come. Lieutenant-Colonel R. D. Johnston was now in command of the Twenty-third, Colonel Christie having been placed in command of General Anderson's brigade.

There is a great gap in the Southern part of the War Records covering the first Maryland campaign. The Confederate reports were either lost or destroyed in that fortnight of strenuous marching and fighting. The casualties of the regiment at Sharpsburg, as at South Mountain, will never be accurately known. Captain Wall's estimate of about 45 wounded and 20 killed is believed to be right. Captain A. T. Cole, Company D, and Captain Wesley Hedspeth, Company E, are the only two officers given in Moore's Roster as having been wounded, though there were almost certain twice or three times that many. Few soldiers in any war have ever been killed under the same circumstances as W. C. Watkins, of Company A. This man had been discharged as not physically able to serve. But wishing to take part in one more battle, he remained and fought at Sharpsburg, and fell and was found dead with the discharge in his pocket.

THE RETURN TO VIRGINIA.

McClellan's desperate and repeated attempts to pierce and shatter the Confederate lines, had been substantially foiled. But Sharpsburg proved to us but a pyrrhic victory at best. Lee with less than 30,000, could not afford victories bought at the expense of 10,000 men, even if it inflicted a loss of 15,000 on the enemy. Holding his lines undisturbed through the 18th, he withdrew that night across the Potomac, near Shepherdstown. Just as the last of our own army crossed the enemy appeared and a brush occurred, but they did not press us closely till the next day, when we turned and drove them back with fearful loss.

After returning to Virginia, our command lay encamped till late in October along the Opequon, not far from its battle ground of 19 September, 1864. The region was one of great thrift and plenty. The long rest was exceedingly grateful to our weary and foot-sore men. During this campaign an intrepid deed was performed near Bunker Hill by Frank Bowers, of Company A. He was then driving an ammunition wagon drawn by six mules. The jolting over the rough road exploded one of the shells in his wagon and others rapidly followed. Few men would have hesitated at instant flight.

No man could have been expected to do otherwise. But Bowers was one of that heroic mold which never abandons a trust or a duty. With marvellous presence of mind and courage, he sprang to the ground, unhitched the team, and escaped with them all unhurt from the verge of the volcano of bursting shells. Yet history vouchsafes this gallant fellow but the stint of two words, one of them abbreviated to a single letter. "k, Gettysburg." (Killed at Gettysburg.)

Here the army was recruited and reorganized. The Twenty-third received its share of recruits. What was more important, it was strengthened by the return of many of its members who had recovered from wounds and diseases. Colonel Alfred Iverson, of the Twentieth North Carolina, was, after Sharpsburg, commissioned Brigadier-General and assumed command of the brigade. The Thirteenth Regiment was about this time transferred to Scales' Brigade, leaving brigaded with us the Fifth, Twelfth and Twentieth

In November came the march southeast to Fredericksburg. The following incident—a trifling flotsam of memory—occurring in this month, will illustrate the humorous side of a soldier's life. One of the Staff officers of the regiment, for slightly overstaying a leave to visit some ladies was, as was the usage, placed under arrest by Colonel Christie. Now an officer under arrest must march in the rear of the regiment, and cannot address his superior officer except in writing. This incompetency to address the Colonel would have been without complications except for the fact that the weather was cold and the above officer and the Colonel were bed-fellows and slept on a very narrow bunk. Now not even a Confederate soldier was willing that all the freezing that fell to his lot should be endured by one half of his body. So an occasional turning of the frozen side in was a *sine qua non*. But a luxury of this kind could be safely obtained only by co-operation—there must be a simultaneous action of both occupants of the bunk or dire consequence might follow. For co-operation communication is essential. Written communication in the dark was impossible. Finally after long consultation with two other officers in the same tent—the Colonel remaining a silent, but doubtless highly amused auditor—it was decided that an officer under arrest might in extremity, address his superior by proxy. This was forthwith done, a change of base effected and Confederate comfort assured.

FREDERICKSBURG.

The Twenty-third took no active part in repelling the Federal army—now under Burnside—at Fredericksburg.

We were held in reserve near Hamilton's Crossing behind Early on the right. Here, though exposed to the artillery fire from Stafford Heights, only one man was killed and a few hit. But Sunday morning, 14 December, our division was carried around and placed in the front line on the extreme right. During the day we affiliated for a while with the Federal officers in our front, truce being granted by Lee to Burnside to bury his dead.

That evening preparations were made for a night attack. A white band on the arm was to be the distinguishing badge of our troops in the night assault. These were provided and, we believe, in a few instances, actually put on. No attack was ordered, the crushing blow that we had so easily dealt the enemy not being yet realized by our commanders.

On Monday night, 15 December, a picket line from our regiment was thrown well to the front. Captain H. G. Turner, of Company H, in command of the pickets, seems to have been the first man in the army to discover signs of the Federal retreat across the Rappahannock. The night was boisterous, a strong northwesterly wind had, as is so often the case, followed the snow fall of some days ago. This wind muffled any sounds in the enemy's lines, which were to the east of us. But Captain Turner observed a scarcely preceptible, though incessant flickering of the lights on the distant hills across the river. This he could account for only on the theory that long columns of troops were there moving under cover of night. This movement he at once construed to be a retreat. What he had seen and the inferences he drew therefrom, were at once reported to his superiors. Nothing came of his report. Soon after Fredericksburg, General Rodes was placed in command of our division, General Hill being assigned to another position.

The battle over, we went into winter quarters near Fredericksburg, out towards Guinea Station. Here, in January or February, 1863, we took part in a great snow battle. The long roll was beaten and the brigade ignorant of what it was to do, fell into line, officers at their posts as if for real battle. Orders were given and we marched rapidly out towards Dole's Georgia brigade, which we were to attack. The Georgians had thrown up breastworks of snow, prepared a supply of snow balls and were ready for us. It was a grateful relief from the tedium of camp life and the men entered with zest into the sport. After preparing as much ammunition as we could conveniently carry, our line moved forward to the assault. The battle, though brief, was sharp, many of us were knocked down and several quite seriously hurt, but the snow fort was stormed, our opponents routed and chased back through their camp. Many prisoners were taken. The horse play was ended by rolling in the snow a supercilious general officer participating in the fun. The irate General sought a court-martial, but was told that an officer waived his rank when he took part in frolics of that kind.

The enemy was still in full force across the river opposite us. This kept Lee's army constantly on the *qui vive*. Our regiment did a great deal of shivering picket duty on the Rappahannock below Fredericksburg. The winter was one of great rigor. The men, though pretty well hardened, suffered severely from want of proper clothing and food and from exposure. Some time in January or February the command was marched to Mine Run, and though they did only a little desultory fighting, they suffered much hardship from cold, being held in line in the snow for several days and nights. The enemy being in sight, no fires could be allowed and our suffering was intense.

THE BATTLE OF CHANCELLORSVILLE.

If the consensus of the intelligent opinion of the world was taken as to what battle gave most lustre to Southern generalship, it would almost surely designate Chancellorsville. Lee holding strong positions along the Rappahannock and higher up upon its tributary, the Rapidan, had deemed it expedient to detach Longstreet to spend part of the winter near Suffolk, Va., gathering supplies from that region and from Eastern North Carolina. Thus less than 55,000 men were left to confront Hooker, who had superceded Burnside as commander of the Federal army.

Hooker took advanage of this separation of the Confederate forces by strategic stroke that may, in its inception, be called brilliant. Making a feint against Lee's front at Fredericksburg and his right below that place, he suddenly 29 and 30 April, 1863, threw 120,000 men across the Rapidan on the Confederate left flank. Had Hooker possessed the hardihood and moral courage of Grant and have advanced from the Wilderness into the open country where his vastly superior force could have told, things must have gone hard with Lee. But as has been well said, while Hooker hesitated, Lee acted. Jackson, with 22,000 men, by a rapid march whose very boldness bewildered the enemy, swept from Hooker's left flank across his front and fell upon the unsuspecting right flank like a bolt from the skies.

The Twenty-third took a highly important part in this brilliant movement. It led the van in Jackson's immortal march. Friday evening and Saturday morning, 2 May, its skirmish line was in contact with the enemy not far from the Chancellor House. At daybreak, it was so hastily withdrawn that two of its companies, then on the skirmish line, were left behind and did not rejoin the regiment till late in the evening.

Our regiment on being withdrawn from contact with the Federals, went swiftly forward through the Wilderness, striking now and then a dim path or road. Strict silence was enforced, the men being allowed to speak only in whispers. Occasionally a courier would spur his tired horse past us as we twisted through the brush. For hours at the time we neither saw or heard anything. Great was the curiosity to know where we were going and what "Old Jack" was about. But we agreed that he did know and that the novel march meant much. Our brigade led the division, our regiment the brigade. While swinging onward a turn in the dim road brought us suddenly face to face with a piece of Federal artillery, which firing point blank, double-shotted with canister, struck down the head of the column, disconcerting for a moment many of the bravest. Major C. C. Blacknall, with rare presence of mind, instantly rallied a company and springing forward with the bayonets, captured the piece before it could be reloaded.

During the afternoon we reached the position assigned to us. The Twenty-third was the very last regiment on the left wing of the army. Tired, breathless, but bouyant, we lay down in the woods near the unwary foe and waited till ordered to attack. As the afternoon passed we were swung around still farther to the left and to the rear of the right flank of the Federal Eleventh Corps. The attack was begun back to our right. As the sun was round and red and low, the regiment moved directly towards it on the foe. At the first sight of the Federals, we were ordered to yell our loudest and to move forward up the hill at the double quick. We struck their very rear, charging in over their beef slaughtering and cooking detail. The enemy began jumping up before us and holding up their hands to surrender. But little resistance was met with, the surprised enemy surrendering or breaking before us in the wildest rout and disorder. Chasing them like hares, our boys surged forward. Prisoners, pieces of artillery, a regimental flag and countless stands of small arms were taken by the Twenty-third. Albutress Gabriel, a private in Company K, captured a brigade commander. The frenzied flight of the foe is well illustrated by a cannon which was seen hanging up a tree. In the panic it had been driven over a small tree which bent under its weight, but finally broke it loose from the caisson in front. Then the upspring of the tree raised the entangled gun from the ground. There it hung as eloquent an attest of mad flight as perchance war has ever seen. We soon emerged into a large field occupied by a large part of Hooker's army. Their line of battle was snugly intrenched, but the works faced the wrong way. We came up obliquely behind their works. Their line, in hurriedly trying to face about and meet us, was soon tangled and scattered pell-mell all over the big field. Over this field rushed helter-skelter cannon, wagons, loose horses, dogs, men, everything. A spectacled Adjutant was here shot dead with a congratulatory order in his hand, telling that Lee was surrounded and would be captured the next day. It was well into the night before our onset spent itself and we must have been then not very far from the point from which we set out in the early morning. Then were heard all through the woods the Yankee officers calling out and offering to surrender. We heard distinctly without knowing its fateful meaning, the sudden outburst of musketry which struck down the right arm of Lee and of the Confederacy—Stonewall Jackson.

Our loss that night was small, as it had been with us rather a chase than a fight. Our turn to fight came the next morning.

On Saturday night both Lee and Hooker made different dispositions for the stern Sabbath work to come. Major Rowe, of the Twelfth, having been wounded the evening before, Lieutenant-Colonel R. D. Johnston, of the Twenty-third, commanded the Twelfth in this battle. Iverson's brigade went into action on the left of the Confederate line and to the left of the plank road. Having been in the first line the day before, it was now placed in the second line as a support. Our brigade reached the first line as it was falling back from its assault on the third line of Federal intrenchments. General Rodes says of this attack: "The enemy was compelled to fall back and pressing on Colonel Hall's two regiments (Fifth and Twenty-sixth Alabama) together with the Twenty-third North Carolina, Colonel Christie, carried the heights in magnificent style, planting their flags inside the works."

The rest of Rodes', Iverson's and Pender's troops were repulsed. This exposed the three above regiments, and an overwhelming flanking attack by the Federal Generals, French and later Humphreys, being made, they were forced to retire with heavy loss in killed, wounded and prisoners. But the troops which had been repulsed soon rallied and on being reinforced, drove back the attacking forces and the general Confederate advance followed.

Major N. A. Gregory (then Lieutenant Company I) gives a graphic account of several incidents in the battle. He says substantially as follows: "They (Pender's men) had captured two lines of works from the enemy and were standing behind the second line when we came up. They told us that they were out of ammunition and could go no further. General Pender went forward with us. After crossing a little branch and fighting for some time in a hot place, Pender told us to charge. We rushed ahead. My company was on the right. I bore to the right of the road and got into a little fort, which stood in the open field near the road. Here I seized a rifle from a man who went into the fort with me and blazed away at the colors of the Federal artillery company that was then moving off the field. Just then this man called my attention to the shots coming in on our left. As we two were alone, we got out of there. I suppose that we went closer to the Chancellor House than any other command that day. These shots were from French's flanking force about to strike the Confederate left."

The loss of the Twenty-third at Chancellorsville, which is said to have been 50 per cent. larger than any other regiment in the brigade, was officially reported by General Rodes at 173 killed, wounded and missing. Moore's Roster gives the casualties as follows: Wounded 48, killed 17, mortally wounded 6. Captain Wesley Hedspeth, Company B, was killed. Lieutenant James S. Knight, Company B, was mortally wounded, dying that night. Lieutenant Washington F. Overton, Company G, was wounded and burned with many of our dead, and probably some other wounded, in the fire that raged that morning in the woods to the left of the plank road and east of the little aldertangled branch. Captain A. T. Cole was wounded and captured while being carried to the rear. Major C. C. Blacknall and Lieutenant George B. Bullock were, with the men they led forward, surrounded and captured in a redoubt of the work which they had just carried. These two officers after being fellow prisoners in the

Old Capitol Prison at Washington with Miss Belle Boyd, the famous Confederate spy, were exchanged in two weeks and took part in the Gettysburg campaign.

THE GETTYSBURG CAMPAIGN.

In no period of the war was the Southern heart more buoyant or did hope gleam brighter or larger than when it was known that Lee's victorious army had invaded the North. All things now seemed possible. But at no period of the struggle was hope really more fallacious and deceptive. Southern Independence had already been lost. Chancellorsville was its grave. With Gettysburg won and Vicksburg lost, Southern Independence could not have been attained. But Chancellorsville won—decisively and overwhelmingly won—Lee could easily have detached a force to relieve Vicksburg. Chancellorsville must have been a decisive and overwhelming victory but for the fatal blunder of one man—a man brave and otherwise competent. At 5:30 p. m. Saturaay evening, 2 May, 1863, Jackson held the fate of Hooker's army in the hollow of his hand. His subordinates had but to move forward when and where he had distinctly ordered, and within an hour a blow would have been struck the enemy, which, followed up with a tithe of Jackson's energy, could have ended only in Hooker's undoing. This unfortunate officer was General Colquitt, commanding a Georgia brigade, to whom had been assigned an exceedingly important position on Jackson's right. The duty assigned this wing was to strike the routed Eleventh Corps on the flank and rear and not only destroy or capture it, but what was even more important, assail the other commands then open to attack. But this duty was never performed. Colquitt saw some horsemen in Federal uniform on his right front. The apprehension of an attack on this flank—an impossible thing—suddenly overcame him. He halted his regiments and changed front and also forced Ramseur's brigade to do likewise. Having the right of way over the "Stonewall" brigade and four regiments of Stuart's cavalry his halt halted them. This change of front and the purposeless marching it entailed, kept inactive seventeen regiments of excellent troops for an hour—an hour as big with fate as battle field ever saw. For in that hour the torrent of Federal rout passed by to a place of safety. Hooker, or his subordinates, made new dispositions and brought up their powerful artillery. When at length the seventeen regiments came up and the Confederates moved forward the golden opportunity had passed; rout and disorder had with the foe given place to order and determination. Jackson, realizing the exigency of the new turn in the battle, went forward to inform himself and fell.

But to return to the Gettysburg campaign. Leaving the vicinity of Fredericksburg 4 June, 1863, we marched, *via* Culpepper Court House and Front Royal, to Berryville, Va. Here the army captured the camp equipage of 1,500 men who fled without a battle. Thence to Winchester, where 3,000 of Milroy's men were taken and marched past our command. At Martinsburg we cut the Baltimore and Ohio Railroad and pressing forward, waded the Potomac at Williamsport on Monday, 15 June. Passing through Chambersburg we reached Carlisle, the northern limit of our invasion, about 27 June. The Twenty-third acted as provost guard at several places on this march. At Carlisle we rested for several days in the Federal barracks. Here many of our jaded, weary boys, drank too much United States Government whiskey and a battle with a Georgia regiment, for the time likewise drowning their weariness, was narrowly averted. Many of the Carlisle people knew General Iverson, he having been quartered in the barracks there when a Lieutenant in the Federal army.

As Lee threw our corps (Ewell's) north to Carlisle, threatening Harrisburg, the capital of Pennsylvania, he concentrated his other two corps, A. P. Hill's and Longstreet's, at Chambersburg on his line of communications. Stuart having taken his cavalry on his famous, but fatal, raid around the Federal army, Lee was long in complete ignorance of the enemies whereabouts.

Orders had already been given for the march on Harrisburg, when on the night of the 29th Lee, then at Chambersburg, learned from a scout that the enemy were on his right flank, the head of their column being then near Frederick, our resting place of the fall before.

Our corps was at once put in rapid motion southward. The intelligence received had changed Lee's whole plan. His plan now was to concentrate at Cashtown with the mountains at his back and beyond them the rich Cumberland Valley, for a granary. Here Meade, who had now superceded Hooker, would have had to attack us with everything in our favor. A. P. Hill, contrary to orders, precipitated battle at Gettysburg with the enemy on the defensive and everything in their favor. However, it is but fair to General Hill to add, that owing to the absence of cavalry, he had no means of knowing that the forces unexpectedly interposed between him and Gettysburg, whither part of his command was marching to procure a supply of shoes, were other than militia or at most a small detachment of Meade's army.

Leaving Carlisle on Tuesday, the last day of June, we marched swiftly southward. Cherries were ripe along the rock-walled lanes. Bringing camp hatchets out, fruit ladened limbs were severed and we regaled ourselves as we swung onward. The spirit and morale of the army were then superb. Many German-descended members of our regiment belonging to the companies raised in Lincoln, Catawba, Gaston and Montgomery Counties, were in this region amid, or not far from, their kin. From here their ancestors had emigrated to North Carolina about one hundred years before. But I doubt if many of them thought of it at that time. Little did the families at the separation imagine that the descendants of the emigrants should in a generation or two return as invaders to the old home. To this day (1900) there are German families around Gettysburg which recognize their distant kinship to and occasionally visit their people who came to this State about 1750.

But to return to the subject in hand. Sounds of strenuous battle reached us early on the morning of Wednesday, 1 July, as we pressed forward towards Gettysburg, the obscure Dutch town so soon to be made famous. Our brigade (Iverson's) led Ewell's corps and was the first to become engaged as he hurried forward to succor A. P. Hill, then hard pressed. At Willoughby Run our Field Officers dismounted. Approaching from the north by the Heidelburg road till within about a mile of the field of battle, we were filed off by the right flank to the Mummersburg road. As we emerged from the woods and moved down the slope to the latter road twenty pieces of artillery opened on us with grape, from the left, inflicting some loss.

The Mummersburg road here runs east and west. Very close to the road on the south side stands the Forney house. This house stands in the northwest corner of the Forney field, which extends about half a mile from the house along the Mummersburg road, and is about a quarter of a mile broad. Across this road near the Forney house the brigade was formed facing east. Along the path or eastern side of the field and on a ridge ran a stone fence, which formed part of the enemy's line. Behind this fence, alone, lay hidden from view, more men than our assaulting column contained. A body of woods extended from the southeastern corner of the field for about two hundred yards along its southern side.

The brigade, about 1,450 strong, advanced under artillery fire through the open grass field in gallant style, as evenly as if on parade. But our brigade commander (Iverson) after ordering us forward, did not follow us in that advance, and our alignment soon became false. There seems to have been utter ignorance of the force crouching behind the stone wall. For our brigade to have assailed such a stronghold thus held, would have been a desperate undertaking. To advance southeast against the enemy, visible in the woods at that corner of the field, exposing our left flank to an enfilading fire from the stronghold was fatal. Yet this is just what we did. And unwarned, unled as a brigade, went forward Iverson's deserted band to its doom. Deep and long must the desolate homes and orphan children of North Carolina rue the rashness of that hour.

When we were in point blank range the dense line of the enemy rose from its protected lair and poured into us a withering fire from the front and both flanks. For Battle's brigade, ordered to protect our left flank, had been thrown into confusion by the twenty pieces of artillery and repulsed by the right wing of the Federal line just as we came up. This effected, the enemy moving under cover of the ridge and woods, disposed his forces to enfilade our right from the woods just as our left was enfiladed from the stone fence.

Pressing forward with heavy loss under deadly fire our regiment, which was the second from the right, reached a hollow or low place, running irregularly north, east and southwest through the field. We were then about eighty yards from the stone fence to the left and somewhat further from the woods to the right, from both of which, as well as from the more distant corner of the field in our front, poured down upon us a pitiless rifle fire.

Unable to advance, unwilling to retreat, the brigade lay down in this hollow or depression in the field and fought as best it could. Terrible was the loss sustained, our regiment losing the heaviest of all in killed, as from its position in line the cross enfilading fire seems to have been the hottest just where it lay. Major C. C. Blacknall was shot through the mouth and neck before the advance was checked. Lieutenant-Colonel R. D. Johnson was desperately, and Colonel D. H. Christie mortally wounded, as the line lay in the bloody hollow. There, too, fell every commissioned officer save one; the recorded death-roll footing up 54 killed and 82 wounded. The real loss was far greater, almost surely 50 per cent. greater. Captain G. T. Baskerville, Company I; Lieutenant C. W. Champion, Company G, and Adjutant Junius B. French, were killed. Captain A. D. Peace, Company E, and Lieutenant Wm. M. Mundy were wounded. Captain H. G. Turner, Company H, was wounded and captured. Captain Wm. H. Johnston, Company K, was captured.

The carnage was great along our whole line which, except the Twelfth Regiment on the right, was at the mercy of the enemy. The Twelfth, under Colonel Davis, protected somewhat by the lay of the field and being further from the stone wall, refused both wings and fighting to right, left and front, gallantly beat off its assailants till help came.

Ramseur was now hastening to our relief. The wary foe aware of this, swarmed over the wall and rushed down upon our weakened line. Leaving the wounded they drove off with bayonets and clubbed muskets 49 prisoners and carried our flag with them.   The One Hundred and First New York regiment has marked with a stone the point reached in this charge.   It stands about where the Twenty-third lay.   This rush was all over in a moment, for Ramseur was coming up. This gallant officer, had he continued to advance as he started, straight against the stone fence, must have met with disaster just as we did.   It is said that Lieutenant Crowder, of Company A, and Lieutenant Dugger, of another regiment, ran back and advised him to file off to the left and strike the Federal right.   At any rate he effected this movement with brilliant and decisive success.   The enemy saw it and apprehending its meaning, strove to change front to meet him. They were too late.   Ramseur caught them in the act, and his rifles silent till then, enfiladed their line along the stone fence with terrible and crushing effect.   This fire also killed Rial Stewart, and perhaps others of our regiment, who had just been captured and were being taken to the Federal rear.

Ramseur's onset began the enemy's reverses which ended in their being driven back through the town of Gettysburg with the loss of 5,000 prisoners, besides many killed and wounded.   What was left of our regiment and brigade went forward in the attack and pursuit.   Fire was opened on us from the houses as we rushed into the place, but we shouted that we would burn the town unless it stopped.   The firing ceased.

General Rodes said that Iverson's men fought and died like heroes.   When the brigade went from its position in the hollow its dead and wounded lay in distinctly marked line of battle from one end to the other.   The imperfect returns show 512 killed and wounded.   The most careful estimate makes it over 750.   A member of the Twenty-third lying stone dead, his musket clinched in his hand and five bullets through his head attests the close and deadly fire under which they lay.   Thirty-five years after the battle the writer found in the clay of the pits from which Iverson's dead had been removed to Richmond, flattened bullets which had evidently fallen from the disinterred skeletons. The field was even then a veritable mine of war relics—bullets, grape shot and pieces of shrapnel. Lieutenant George B. Bullock, Company I, said that it was the only battle—and he was in all in which his command was engaged from Williamsburg to Appomattox—where the blood ran like a branch. And that too, on the hot, parched ground.

The handful left of our regiment were not taken into action on the second or third day at Gettysburg. While being conveyed, wounded, on the retreat through South Mountain Sunday night, Colonel Christie, Lieutenant-Colonel Johnston and Major Blacknall were captured by Kilpatrick's cavalry near Monterey Springs. Christie and Johnston were rescued by the Confederate cavalry and carried to Williamsport. Blacknall escaped on Kilpatrick's own horse, but being too badly wounded for rapid flight, was recaptured.

Colonel Christie died at Winchester, Va., and in his native county, soon after the army reached that place. His wife, whom he so longed to see, and who had hastened to him, arrived a few hours after he was buried.

COLONEL D. H. CHRISTIE.

Daniel Harvey Christie was born in Frederick County, Virginia, 28 March, 1833, and was educated at a military school. He became a citizen of Henderson, N. C., in 1857. The breaking out of the war found him in charge of the Henderson Military Institute which he had established. His gallant conduct and wounds at Seven Pines and Cold Harbor have already been mentioned.

Although the latter wound was very severe, within sixty days he returned to his command and devoted himself diligently to the work of recruiting and disciplining his regiment. At South Mountain his management of his regiment was such as to elicit from General Garland words of the highest praise for himself and his regiment, a few minutes before Garland fell. After Sharpsburg he commanded Anderson's brigade till Colonel Bryan Grimes reported for duty. At Gettysburg, his last battle, Christie's conduct was especially gallant. Here he held his men in position under a most terrific fire for an hour till the whole regiment was killed, wounded or captured, except a Lieutenant and sixteen men. He was in the act of leading a charge against the stone fence when he fell, with his men and officers thick around him. Colonel Christie was buried at Winchester, another Colonel of the Twenty-third being laid by his side a year later.

CAPTAIN BASKERVILLE.

George Thomas Baskerville was born in Mecklenburg County, Virginia, 16 October, 1827. He graduated with high honors at the University of North Carolina at the age of 17, being the valedictorian of his class—delivering his address in Latin. About 1849, he became a citizen of Granville County.

Captain Baskerville was without military ambition. But, impelled by a strong sense of duty, he joined the army and was elected Captain of Company I, Twenty-third North Carolina, in 1862. Refusing promotion he remained with his company, serving with courage and ability. Falling, wounded to death, at Gettysburg, he died the next day. His devoted wife crushed at the tidings of his death, took to her bed and never rose again. Captain Baskerville was of the highest type of Southern gentlemen. He was a devout Christian, a good neighbor and a devoted husband. His domestic life was a most beautiful one. Plighting their troth when children, marrying very early in life, their devotion to each other was complete. And when the sturdy oak was stricken down, the clinging vine fell with it.

The virtual destruction of Iverson's brigade at Gettysburg was largely, if not wholly, owing to the fact that it had no brigade commander on the field to govern its movements, as a whole, in accordance with the exigencies of the battle and to halt it before it entered, unsuspecting, the deathtrap laid for it. Iverson's part in the heroic struggle of his brigade seems to have begun and ended with the order to move forward and "Give them hell." The brigade refusing to serve under him longer, he was transferred to the cavalry and Lieutenant-Colonel R. D. Johnston was commissioned brigadier and assumed command on 8 September, 1863. General Iverson's conduct at Chancellorsville had also been severely criticised. Where he was when Pender led forward his (Iverson's) brigade, has never been explained. The Confederate newspapers of that period spoke of strained relations between Lee and Davis because Davis refused to let Lee court-martial the "delinquent brigadiers" for their action, or rather non-action, at Gettysburg. However, the fact of any coldness between them was denied.

THE RETURN TO VIRGINIA.

On the retreat we crossed the Potomac at Falling Waters near Williamsport, 10 July. After operating in the valley for a short while, our corps moved towards Madison Court House. Here we rested till Lee's move 9 October to strike Meade's flank, who was then at Culpepper Court House. On that march the Twenty-third, Fifth and part of the Twelfth, all under command of Lieutenant-Colonel Davis, of the Twelfth, crossed the Rapidan at Raccoon Ford. This detachment was highly commended for a gallant charge on a battery and its support. The regiment sustained loss both at Vidiersville and near Brandy Station during the same movement.

We went into winter quarters near Orange Court House. But, in February, or March, our brigade was detached to guard bridges over the North and South Anna rivers, near Hanover Court House. Here we were recruited, equipped, and put in good trim.

In barracks at Taylorsville, near Hanover, with no enemy near, the command had the only really good time during the war. The only thing like work was the attempt to overtake the raiding force under Dahlgren. Neat uniforms and even pleated-bosom shirts, long unknown, were here to be seen, and some of the boys bent on luxury in the extreme—thoroughgoing sybarites—actually boarded out. Eating regularly three times a day, keeping dry and sleeping warm of nights seemed a preposterous thing to a Confederate soldier. We even went into politics. 11 March, 1864, the brigade held a convention at Taylorsville, endorsing Vance as against Holden and his treasonable influences. But there never was a

100   **NORTH CAROLINA - 23rd REGIMENT INFANTRY - Roster and History**

****

dream so bright, or paradise so sweet that some one did not come to spoil it. Grant spoiled ours.

Beginning at midnight of 3 May, 1864, Grant, now Commander-in-Chief of the Federal armies, suddenly threw nearly 120,000 men under Meade, across the Rapidan. Grant's plan was to flank Lee out of his entrenched position on Mine Run and fight him somewhere between the river and Richmond if he would stand. That Lee did stand is attested by the fall during this movement of more Federals than Lee had men. Our brigade left Taylorsville at 11 a. m., 4 May and by the quickest forced march on record covered sixty-six miles in twenty-three hours. Army mules fell dead in their traces under the severe strain, but without stopping for bivouac, or hardly for rest, we held out and reached the plank road near the Wilderness Tavern, on the 5th.

Dead tired as we were, we were ordered forward about sunset, with J. B. Gordon's brigade. The movement was under Gordon's command and was directed against the Federal right. Driving the enemy back a mile or more with slight loss to ourselves, we halted on the turnpike and slept as even tired soldiers hardly ever slept before. During the night of 7 May, Grant began his flanking movement around Lee's right. Lee swung Anderson's division around and headed him off at Spottsylvania Court House. It was while on the march to Spottsylvania that Johnston's brigade was, much to their regret, transferred from Rodes' division to Early's, Early being assigned temporarily to the command of Hill's corps, Gordon commanded the division.

On the 9th, at Spottsylvania, our brigade, with 300 or 400 men, made a reconnoissance on the Confederate right and drove back a division of Burnside's corps, but seeing himself nearly enveloped by the enemy in overwhelming force, Johnston withdrew his brigade in time to escape capture. The Twenty-third lost .20 to 30 men in this move. Sergeant Thomas Powell was wounded, captured and died a few days later in Washington. Lieutenant-Colonel Davis, of the Twelfth, was now in command of the Twenty-third.

About this time the brigade now but a handful, fought and ran off a heavy cavalry force endeavoring to hold the high and open ground around the old court house at Spottsylvania. The cavalry was a splendid body and fought desperately, but no incident of the war was more relished by the boys than trouncing and chasing that prim set of blue-coated horsemen.

We took no part in the battle of 10 May till nearly sunset. The enemy had broken over the works by Dole's brigade and were advancing direct against General Lee's headquarters. While other troops assaulted his flanks, our brigade took the most important part in repelling this assault of the enemy. The men refused to go forward till General Lee, then on the field, went to the rear. The following account of the battle is from notes of Captain A. T. Cole, made not long after the war: "About sunset the enemy broke through our line at an angle in the works and were advancing rapidly towards General Lee's headquarters then in sight, and directly before them. Our brigade was doubled-quicked by the right flank in column from behind a pine thicket where it had been resting and concealed. Emerging suddenly in their front, then going by the left flank in line of battle, we met and drove the enemy back across the breastworks and regained several pieces of artillery which were still in position. Some of the Confederate gunners who, concealed in the cannon pits, had escaped capture, now sprung out and used the guns very effectively on the retreating Federals. Just as the brigade faced by the left flank and advanced towards the enemy, I saw facing the head of the column General Lee on horseback, hat in hand, cheering on the men, within not more than 100 yards of the enemy. The fighting lasted till probably 9 o'clock that night. Killed and wounded in our regiment numbered 20 to 25."

In making the charge Major Brooks, of the Twentieth North Carolina, and Captain Jos. F. Johnston, Aid-de-Camp to General R. D. Johnston, were competitors in a race for a Federal flag which had been planted on the captured Confederate works now held by three lines of battle. Brooks reached out his hand just in front of Johnston and seized the flag, carried it back to the rear and presented it to General Lee with the request that it be sent back to North Carolina as one of the trophies of the brigade. It was sent to this State with a letter from General Lee very complimentary to North Carolina troops.

After repulsing the attack of the 10th, the brigade was again withdrawn, occupying its place on reserve till the 12th. Daybreak 12 May, a foggy, dismal dawn as May ever saw, found us at the Harris House half a mile to the rear of the apex of Lee's salient, thence forward to be known as the "Bloody Angle." The Confederate line of fortification swept around Spottsylvania Court House in an irregular semi-circle seven miles long. A mile due north of the Court House a spur in the hills made it necessary, in order to prevent the enemy from occupying a commanding position, to construct a great angle or salient in the works. This salient, not unlike a huge horse shoe in shape, was about three-fourths of a mile long and half a mile broad at its base. This position, with artillery, was strong; but without, it was weak. Lee believing that Grant had resumed his movement by the left flank, had ordered the withdrawal of all artillery on this part of the line not easy of access. On the night of the 11th General Ed. Johnson, who with his division of 2,000 men, held the toe of the horse shoe, apprehending an attack from the movements in his front, asked that the artillery be returned. The guns were just going back into position when at daylight Grant threw a solid mass of 20,000 men against Johnston's 2,000, taking the guns before they could open fire. The victorious enemy then pressed onwards to seize the whole salient and pierce Lee's centre. Our weak brigade in bivouac at the Harris House, half a mile to the rear, were the only troops immediately available to stem the onset. General Gordon at the sudden outbreak of battle, threw us forward. Going forward at the double quick in the woods below the McCool House and far down in the salient, we ran upon the Federals coming forward in three dense lines of battle. Our numbers were so few and the enemy so strong, the intervening distance so short, that twice Federal Line Officers came within ten steps of us and demanded the surrender of the brigade. Our reply in both instances was a volley that struck down the ven-

NORTH CAROLINA - 23rd REGIMENT INFANTRY - Roster and History   103

****

turesome officer and for a moment staggered the oncoming host.

But what availed a few hundred against 20,000. The brigade after one of the bloodiest combats of the war and with heavy loss, was forced backward fighting desperately as it went. Other troops soon came up, striking the invaders on both flanks. The brigade reformed and renewed the battle. General R. D. Johnston seized the flag of the Twenty-third and ordered a charge. The brigade rushed forward carrying the position in their front, Johnston falling wounded as he planted the flag on their works. The struggle continued with the utmost fury till night. On the 14th Lee withdrew to a line of works constructed across the base of the salient. Our regiment, though small, contained many a gallant spirit and many heroic deeds were done on that dark and dismal morn. E. S. (Scip.) Hart, the flag bearer of the Twenty-third, was especially brave; again and again rushing forward with the colors, which were never for a moment lowered except when Scip was felled by a clubbed musket in the hands of a stalwart Yankee. Among the captured on that terrible day was Captain A. D. Cole, Company A. It was Captain Cole's fate, along with Lieutenants Coggin and Bullock, to form three of the six hundred officers which the Federals placed under the Confederate fire at Charleston for several weeks in 1864, and to endure the horrible tortures inflicted on them by starvation at Port Royal a little later. The minute stint of spoiled meal—a gill a day—and pickle on which they subsisted for forty days ended by killing Lieutenant Coggin and bringing Captain Cole to death's door and keeping him there for agonizing months and even years. This too, crowning three years of gallant service in the field. To few, if any, of all the sons of the South was it given to endure more and suffer more in her defense than did this gallant officer.

THE MARCH ON WASHINGTON.

After the Bloody Angle fight our brigade was engaged in the battle of the 19th. Colonel C. C. Blacknall, commissioned Colonel 15 August, 1863, had been exchanged in May, and after commanding a brigade at Petersburg for several weeks, joined his regiment about 1 June and assumed command. Leaving camp near Cold Harbor at 3 a. m. 18 June 1864, we went with the corps now commanded by Early, on the Valley campaign. Marching to the railroad at Charlottesville we took cars for Lynchburg, on which Hunter was rapidly advancing. We arrived just in the nick of time to save the town. Passing at double quick through the streets, within twenty minutes after leaving the cars we were skirmishing with Hunter's advance guard.

Lieutenant Crowder, the same officer whose suggestion to General Ramseur at Gettysburg proved of so much value, and a brave and efficient officer, was severely wounded that night whole posting the picket lines.

Skirmishing at Liberty and driving Hunter across to Salem and westward into the mountains, Early wheeled suddenly up the valley.

There was a little loitering to see what Hunter would do, during which the army making a detour crossed the Natural bridge and rested there a few hours, which detour to see the bridge was put to a vote of the men and carried by a small majority. Leg-weariness is a great stifler to curiosity. However, pretty soon the race up the valley begun. Staunton was reached 27 June. Pressing rapidly forward we reached Harper's Ferry on 4 July. Our advance had been so rapid and unexpected that we here surprised and broke up a Fourth of July celebration, our advance guard eating with appetites whetted by hard marching, the feast not intended for us. As the enemy held the heights beyond the river and commanded the approaches to Harper's Ferry with artillery, only the skirmish line went into the town, except a few venturesome officers who galloped down that night, fired on in every moonlit stretch by the Federal guns. Crossing the Potomac a few miles above, our forces for a few days made feints here and there to confuse the enemy as to our designs.

But, finally, we dashed off for Washington. On 9 July we met Lew Wallace at Monocacy Junction, near Frederick, Maryland, who gave battle to bar our way to the Federal Capital. Resisting our advance through the town, Wallace made a determined stand at Monocacy river.

While Gordon's Division crossed the river and struck

the enemy's right flank, Johnston's brigade was ordered to capture a block house on the other side of the Baltimore and Ohio Railroad. A considerable force of the enemy were in a railroad cut and perfectly protected. The Twenty-third under Colonel Blacknall, made a dash for the block house, but were met by a hot enfilading fire from the line of battle in the railroad cut. A heavy battery across the river also swept them with a raking fire. Captain W. C. Wall, Company F, was severely wounded. Colonel Blacknall was stunned for the moment by an impact of a bullet on the head, which fortunately did not penetrate, and the regiment was driven back.

Upon the failure of the Twenty-third to carry the blockhouse, General Johnston ordered Colonel Davis, of the Twelfth, to carry it. Colonel Davis says: "General Johnston was not in a good humor and I was suffering (sick) so that I could hardly walk. However, I went forward to the ravine (not knowing of the cause of the falling back of the Twenty-third) and here halted and had picked men as videttes to reconnoitre and see all they could. Finding about the line of battle on the railroad, I sent General Johnston a message that if I advanced I would expose by men to an enfilade fire and that if he would dislodge the line of battle in the railroad cut, I could take the house without loss of men. I never heard from General Johnston. In the meantime the fight was going on on the other side (of the river) between Wallace, of Ben Hur fame, and Gordon. Three lines of battle engaged Gordon's one, and now General Wallace begins to retreat. His men on our side then had to pass over quickly or be taken. I moved forward, and as we struck the bridge on one side the enemy were clearing it on the other." This rapid retreat of the enemy was also expedited by a company that passed under a culvert and opening a flank fire on the cut, drove the enemy out.

Wallace was defeated, with the loss of 700 prisoners—our casualties being about the same—and thrown back upon Baltimore. The way thus opened we advanced a forced Sunday march on Washington. Hot, jaded and footsore, we arrived in sight of that city and only three miles distant at 3 p. m. on Monday, 11 July. The day was one of overpowering heat. The troops were too completely exhausted with hard marching to have been effective in immediate attack. A reconnoissance was made Tuesday, but the works were too strongly manned for our 10,000 men to carry. In the reconnoissance Melville Holmes, a lad belonging to Company G, of our regiment, is said to have fallen nearer to the works of the Federal Capital than any other Confederate soldier of the war. This is also said to have been the only instance in the history of the country in which a President of the United States appeared on a field of battle. Mr. Lincoln came out to the works on Tuesday to view the situation and a surgeon was shot very close to his side by Confederate sharpshooters.

Our brigade bivouacked in the grove of the famous Blair mansion. Here an 11-inch shell from fort Massachusetts burst in the midst of the officers' mess at noon on the 12th, fortunately with no worse result than knocking the food out of some of their hands. The unauthorized burning of General Blair's house, if done by Confederates at all, was the work of stragglers. Though there is a strong probability that it was ignited by shells from the fort that made our dinner party its target.

Early's division had now effected all that could be expected in drawing troops from Grant's hosts around Richmond. Federal troops were now hastening to close the passes of South Mountain and the fords of the Potomac in his rear. Therefore after maintaining a threatening attitude against Washington all day of the 12th, and driving in a strong reconnoitering force from the works, he retreated at dark without molestation. Passing swiftly through Rockville and Poolsville, we crossed the Potomac at White Ford, near Leesburg, on the morning of the 14th, bringing off all prisoners and captures in safety.

Resting on the 14th and 15th near Leesburg, on the 16th we resumed the march through Snicker's Gap to the valley, the enemy following. Occasionally we had a skirmish with their cavalry.

THE VALLEY CAMPAIGN.

By this time the Federals were in strong force at Harper's Ferry. Moving by Hillsboro in Loudon County, they struck our wagon trains on the 16th, inflicting some damage. We were engaged in the brush that drove them off. On the morning of the 17th, we crossed the Shenandoah at Castleman's Ferry and took position at Berryville, our division with that of Rodes, guarding the Harper's Ferry road. There was skirmishing with the enemy on the Shenandoah. On the night of the 19th our division, Ramseur now in command, was moved back towards Winchester to protect the town from the now aggressive Federals. On the 20th Ramseur moved upon Stephenson depot, near Winchester, to attack Averill. The division while moving by the flank, was suddenly assailed by a large force of Averill's cavalry advancing in line of battle. Thus surprised, the division was thrown into disorder. But Colonel Jackson made a gallant charge with his cavalry and Ramseur rallying his men, Averill was driven off.

The Richmond *Sentinel* printed about this time a communication very disparaging to the North Carolina troops, and especially to Johnston's brigade, exalting Pegram's Virginia brigade at their expense. In a word it was claimed that Johnston's men ran without firing a gun and that Pegram's redoubtables alone saved them from annihilation. Colonel C. C. Blacknall in a letter a few days after the battle, after referring to the false and deprecatory account of the affair as published in *The Sentinel,* says: "The truth of the matter and which will be attested by every gentleman who was present, was as follows: General Ramseur marched the division down the Winchester road and from the reports of the officer commanding our cavalry in front, was led to believe that the enemy in small force were at a point more distant than we found them to be after reaching the body of woods where our cavalry were in line of battle. General Ramseur formed Hoke's Brigade on the left and Johnston's on the right of the road. Pegram being in the rear when we suddenly found the enemy in a field, immediately in our front, we advanced and engaged him without hesitation, our men advancing under a heavy and destructive fire in splendid style. The enemy's line in the meantime overlapping Hoke's left and pouring into his flank a heavy enfilade fire which caused his left regiment to give way, the panic being communicated to the other regiments of the brigade, each one in turn falling back hastily and in some confusion. While this was going on, Johnston's Brigade was steadily advancing, having broken the enemy's line in our front and caused him to fall back before our advancing column. The left of our brigade, the Twelfth and Twenty-third Regiments, had advanced to within sixty yards of the enemy's line of battle, and every man was standing up manfully when our left was suddenly uncovered by the falling back of Hoke's brigade, the enemy pouring in a large force immediately on our flank. Our little brigade being alone and unsupported were, from the nature of the case, compelled to retreat or be captured, as we could not resist the immense odds which were hurled against us.

"Pegram's Brigade being in the rear of Hoke's, joined in the race and made its escape from the place of danger as fast as heels could carry them without even attempting to make a stand. After falling back to the railroad, some distance, it was thought necessary to make a stand to cover the retreat when the Twelfth and Twenty-third North Carolina Regiments, commanded respectively by Lieutenant-Colonel Davis and 'myself were the only troops that could be rallied; all the rest of the division retreating in much disorder to the fortifications. When we marched back to the line where the troops had been halted, we found Pegram's Brigade had gotten there some time before us, although the world has been informed through the papers that they covered our retreat. General Ramseur stated to General Early that 'Johnston's Brigade whipped everything in its front and was last to leave the field,' which is known to be true by every man who was engaged in this unfortunate affair. The enemy had many killed and wounded in our immediate front, which indicated very conclusively that we were not stampeded without firing a gun as these veracious correspondents would make the readers believe."

In Early's victory over Crook at Kernstown, 24 July, our division was sent to the left to get around Crook's right flank at Bartonsville. Crook here occupied exactly the same position that Shields did when Jackson fought him 23 March, 1862. The result of the battle was that Crook was defeated and driven back in great rout.

Then followed much arduous marching and counter-marching to meet and check the strong and active force which was placed under Sheridan's command early in August. The open valley country with its excellent roads gave great facility for the advantageous use of cavalry, in which Sheridan was overwhelmingly strong. Our boys also did much hard work in reaping, threshing and grinding grain for food. This labor could often be done only under the protection of our guns. The Richmond *Examiner* grew facetious over the merry harvesting time Early's men were having in the valley. Colonel Blacknall, writing under date of 28 August, 1864, says: "You have seen, perhaps, some facetious descriptions of our doings and not doing in the Richmond *Examiner*. The descriptions are drawn in the *Examiner's* inimitable style and quite laughable withal to one at a distance. Still the 'frugal swains' and the 'gentle shepherds' have not had quite so gay and festive a time as one might imagine; we have, it is true, been engaged in reaping and thrashing and gathering supplies from the teeming abundance in the country; but the piping and fiddling and feasting and frolicking, exist in the editor's fertile imagination. The lowing and bleating herd are the beef cattle which affords some very tough steak and the meandering, bubbling streams and gentle flowing rivulets are often very muddy pools from which man and mules all drink indiscriminately, neither thinking themselves better than the other. If, however, any gentleman is disposed to believe that this is a gay thing, all I can say to him is, that we have a good opening for any such to come and try it."

THE BATTLE OF WINCHESTER.

The battle of Winchester found our little army in the valley divided.   General Early has been much criticised for allowing his force to be attacked in detail—for "fighting by divisions," as General Lee termed it.   But the broad open valley country with its many roads along which the strong and active Federal cavalry could operate on his communications, prevented that concentration which would have made the Confederate force a unit.   For Early, with 8,000 muskets, 2,500 cavalry and 1,000 artillery had, as best he could, to hold the valley against Sheridan's 35,000 infantry, nearly 10,000 cavalry and an artillery force nearly or quite as large in proportion to his army as Early's was.   Round numbers are given, as the exact numbers are not known, but they are very close.   Sheridan's numbers as given by Judge Montgomery, are considerably below those usually accepted.

Sunday night, 18 September, 1864, found Ramseur's division out on the Berryville Pike east of Winchester.   Johnston's Brigade was in advance with the Twenty-third Regiment, thrown out on picket near the edge of the woods that skirt the Opequon.   Their position was a little north of the pike, but very near to it and a mile or more from the stream. As the enemy was known to be in force just over the creek, the men were told that they now occupied the exact position in which a Georgia Regiment had been captured and were ordered to be on the alert.

The mounted videttes at the ford of the Opequon must have been captured or eluded, for at earliest dawn Sheridan's troopers swarmed up out of the ravine around the advance pickets of the Twenty-third, so quickly that the pickets barely had time to fire before the horsemen were in their midst.   A few minutes later an overwhelming force of cavalry, closely followed by infantry, charged our weak regiment.   Disputing every inch of ground, making stand after stand, we were driven back upon the brigade and that back upon the division. In one of these stops Colonel Blacknall received his mortal wound and was borne back into Winchester.

General Bradley T. Johnston gives the following vivid picture of that gallant twilight combat: "By daylight, the 19th of September, a scared cavalryman of my own command nearly rode over me as I lay sleeping on the grass and reported that the Yankees were advancing with a heavy force of infantry, artillery and cavalry, up the Berryville road. Johnston and I were responsible for keeping Sheridan out of Winchester and protecting the Confederate line of retreat and communication up the valley. In two minutes the command was mounted and moving at a trot across the open fields to the Berryville road and to Johnston's assistance. There was not a fence, nor a tree, nor a bush to obscure view. We could see the crest of a hill, covered with a cloud of cavalry and in front of them—500 yards in front—was a thin grey line moving off in retreat, solidly and in perfect coolness and self-possession. * * A regiment of cavalry would deploy into line and their bugle would sound the "charge" and they'd swoop down on the "thin grey line of North Carolina." The instant the Yankee bugles sounded, North Carolina (Johnston's Brigade) would halt, face by the rear rank, wait until the horses got within 100 yards and then fire as deliberately and coolly as if firing volleys on brigade drill. The cavalry would break and scamper back and North Carolina would "about face" and continue her march in retreat as solemnly and with as much dignity as if marching in review. But we got there just in time—that is to engage cavalry with cavalry, and held Sheridan in check until Johnston had got back to the rest of the infantry and formed in line at right angles to the Pike east of Winchester."

Johnston reached his supports, though with loss, and from then till 10 o'clock Ramseur's weak division of 1,700 men, unaided except by Lomax's and Jackson's cavalry, held the foe at bay. *Bend* this line perforce must, under the onset of Sheridan's immense force, but *break* it did not. At 10 Rodes' division came up and a little later Gordon's. And all through that splendid autumnal day the battle held. History calls it the battle of Winchester. Locally it is known as the battle of Hackwood from the Hackwood farm on which it was fought.

Before noon the Federals were, by a bold assault, driven back in disorder. But it had been at fearful cost. Rodes and many other gallant officers had fallen, and the Confederate forces were too worn out by marching and fighting and too weakened by losses to take full advantage of Sheridan's discomfiture. And a little later a fresh corps were hurled against our necessarily long and attenuated line. Thus, as morning wore to evening, continued the strife. By strenuous and desperate fighting Sheridan's hosts were held at bay in front.

But troops thus beset could not be expected to bear unmoved an attack in the rear. When late in the day two divisions of Federal cavalry drove in the weak force guarding the Martinsburg road and pressed forward to the outskirts of Winchester in the rear of our left, Early's line wavered, broke, and the army were driven back. General Early distinctly says that our division, Ramseur's, fell back on the right in good order, taking position to keep in line with the other troops. Indeed those movements must have been effected with great steadiness for the division was taken for the left wing of the enemy advancing to envelope the Confederate right on which lay the line of retreat and the report came near causing a panic at another part of the line.

Night approached and the Confederate line crumbled under repeated assaults in front and flank. General Early in his memoirs, says: "Nothing was now left for us but to retire through Winchester, and Ramseur's division, which maintained its organization, was moved from the east of the town to the south side of it, and put in position, forming the basis for a new line, while the other troops moved back through the town. * * When the new line was formed the enemy's advance was checked until nightfall and we retired to Newton without serious molestation.

The exact doings of the Twenty-third on that hard foughten field have not been recorded. All that is known is that it stood firmly, fighting manfully among Ramseur's 1,700 heroes.

Lomax had held the enemy's cavalry on the Front Royal road in check and a feeble attempt at pursuit was repulsed by Ramseur near Kernstown. The army retreated that night to Newton. At daylight we moved to Fisher's Hill without molestation.

Colonel Blacknall being too painfully wounded for hasty

NORTH CAROLINA - 23rd REGIMENT INFANTRY - Roster and History   113

****

removal, was left in Winchester, where his wounds terminated fatally.

COLONEL C. C. BLACKNALL.

Charles Christopher Blacknall was born in Granville County, N. C., December 4, 1830.

His grandfather, Thomas Blacknall, of Virginia, was, at 16, a soldier under Washington. His grandfather's grandfather, the "Reverend John Blacknall, Gent'n," though later of Virginia, was one of the first Episcopal clergymen to officiate in North Carolina. Thence through English country gentlemen of record, his lineage runs back to the Blacknalls of Wing, Buckinghamshire, whose armorial bearings were two centuries old when Columbus sailed to discover the new world.

Charles Blacknall was educated for the law, but never practiced. When the war came he promptly raised and was elected Captain of the Granville Rifles, which became Company G, of the Thirteenth, later the Twenty-third North Carolina Regiment. 15 June, 1862, he was commissioned Major and 15 August, 1863, Colonel of the regiment.

His gallantry at Yorktown, Williamsburg, Seven Pines, Chancellorsville and Gettysburg, and his severe wounds at Seven Pines and Gettysburg and his capture, escape, and recapture after the latter battle, have already been told in the body of the sketch. Severe illness contracted while on duty in the Chickahominy swamps prostrated him and kept him out of the Sharpsburg campaign. Only disabling wounds or prison bars kept him from participating in all battles in which his command engaged up to his death.

Few, if any, North Carolinians had a more romantic or eventful military career than the subject of this sketch. Soon after his recapture on the retreat from Gettysburg and while imprisoned at Fort McHenry, near Baltimore, lots were drawn to select a Confederate officer to be hung in retaliation for a Federal officer about to be executed in Richmond as a spy. Colonel Blacknall drew the black bean. Though finally spared, it was only after a long suspense.

Then followed a rigorous imprisonment at Johnston's

Island, Lake Erie, during the severe winter of 1863-'64. Driven to desperation by cold and hunger the eighteen hundred Confederate officers there imprisoned, planned an escape to Canada. Colonel Blacknall, well known to be ever forward in the charge, was elected one of the officers to lead the forlorn hope in the assault with brick-bats against the guards on the wall that encircled the prison. But there was in their midst a Federal spy, disguised as a Confederate officer. Their plans were betrayed and the guards so heavily reinforced, that men even as desperate as they were, could see no hope of success.

His name standing alphabetically near the head of the list, he was paroled in March, 1864, before the cartel was stopped. Exchanged early in May, he started for his command the day that the Federals cut the Weldon road at Stony Creek.

Apprised of this on reaching Weldon, he returned to Kittrell, his home, and without arousing his family, took horse at midnight and hastened to Petersburg. Arriving there, he was placed in command of a brigade, but ordered back to his regiment before it went with Early's force to the Valley.

In all the arduous marching and counter marching, and in the battles and countless skirmishes of this strenuous campaign, Colonel Blacknall took an active part till mortally wounded early on September 19, 1864. On the evening of the 18th, his regiment was placed on outpost duty on the Berryville pike, two or three miles east of Winchester. The writer of this (V. E. Turner) spent that night with him under a simple fly tent. At dawn on the 19th, sharp firing on his advanced picket line told that the expected attack had begun. At this Colonel Blacknall rode hastily to the front to direct his regiment in the encounter. He remained mounted and held his small force pluckily against the heavy advancing columns of the enemy. In the midst of this and while being borne back by overwhelming superiority of numbers, but contesting every inch, Colonel Blacknall received a severe and acutely painful wound in the ankle, and was carried back to Winchester. The surgeons disagreed as to the necessity of amputating the foot to save his life, and his wish to save it was yielded to. It being deemed risky to carry him in the retreat, he was left in Winchester when the army fell back at the close of the day. So much interest was felt in his behalf that the hospital steward was ordered to remain and take care of him.

That was the last seen of the gallant Colonel by his friends of the regiment. The foot was amputated by Federal surgeons, but too late to save his life. Tenderly nursed by the devoted women of Winchester, he lingered for six weeks and six days, dying November 6, 1864. By a singular coincidence death came to him in the house of a Washington (Mrs. Byrd Washington) and on the site of Washington's old fort (Fort Loudon) built in the French Indian War.

Colonel Blacknall was buried by the side of Colonel Christie, his predecessor in command of the regiment—*par nobile fratrum.*

Colonel Blacknall was a man of varied gifts. He loved letters and his reading had been considerable and of the best. He was a strong and graceful writer and a ready and eloquent speaker. To few of the children of men has been given as much personal magnetism. During his three and a half years' service as a soldier no one in the regiment was more beloved; no one behaved more gallantly; no one endured the deprivation and hardships of army life more cheerfully.

Courage was the common staple of Confederate soldierhood. But Charles Blacknall had a command of faculty and an ability to think and act in an emergency possessed by few. One who knew him well spoke of him as one of the few thoroughly chivalrous men that he ever knew; another as the ideal Confederate officer. A chapter of Daughters of the Confederacy at Kittrell, Vance County (formerly a part of Granville County) Colonel Blacknall's home, has been named for him.

General Pegram was now placed in command of our division, Ramseur being placed in command of Rodes' division after the death of that officer. Captain Frank Bennett, Company A, by seniority of rank, assumed command of the Twenty-third on the fall of Colonel Blacknall.

THE FISHER'S HILL DISASTER.

22 September Sheridan, who had followed us and occupied our front in force, threw Crook's corps on the left flank of our line which, even when stretched to the utmost, was too short to occupy the position held.   Driving back Lomax's weak line of dismounted cavalry, Crook advanced against Ramseur's left flank.   Desperate efforts were made to throw Ramseur's brigades and then our division (Pegram's) into line to the left.   But this movement in the face of a vastly superior enemy, could not be effected without disorder. Crook taking advantage of this, advanced, and after a brief contact forced the whole army back in confusion, capturing eleven of Early's guns.

The Confederate foot soldier was not noted for his admiration or his respect for his compatriot who bestrode a horse. Early's foot soldiers' love for a cavalryman was even below the Confederate average.   Sheridan's horse was so much stronger in numbers and equipments than ours, and the nature of the country gave this superiority such opportunity, that our cavalry, gallant fellows as they were, had no showing and cut a poor figure.   But the man who trudged and toted a musket, made none of these allowances for his mounted brothers, who dashed hither and thither with no object apparent to prejudiced eyes, except that of keeping as much space as possible between themselves and the foe.

For some cause known only to their whimsical philosophy, Imboden's cavalry was an especial object of their disesteem. By way of derision they called it "Jimboden's" cavalry. The confidence in General Early had met with that impairment which is almost sure to be the lot of the unsuccessful leader, no matter from what cause.   This spirit in the troops manifested itself at Fisher's Hill in the most drolly humorous incident of the writer's whole war exprience.   Close beside the road along which the troops poured in confusion, a ragged, dejected, unkempt "Confed" crouched over a little fire, regarding naught, absorbed alone in warming numbed fingers and toes, for the day was chilly.   As he crouched and shivered he droned a song in whose tone disgust, despair and disdain all strove for the mastery. The song, which must have been rich, was lost except the following stanzas caught as a group of officers rode by:

"Old Jimboden's gone up the spout,
And Old Jube Early's about played out."

"Gone up the spout" was war lingo for passed into nothingness, even as water in a kettle does when it evaporates and goes up the spout. The singer seems to have believed that Imboden's instead of Lomax's cavalry was the force that proved unable to cope with the enemy on our left flank that morning.

Halting at Mount Jackson on the 23d to enable the sick, wounded, and hospital stores to be carried off, the retreat was resumed to Rude's Hill. Hither the close pursuit and flanking movements of the enemy forced Early to retire in line of battle, a most difficult operation when done under fire and exposed to repeated assaults which had to be beaten off. Nine miles of the retreat was thus covered, the troops passing through the ordeal of repeated attacks with great coolness. While thus fighting and falling back with the steadiness of Cæsar's cohorts, by a strange coincidence we came to a place called "The Tenth Legion." Here at sunset we made a stand and checked the pursuit for the night.

Retreating up the valley, constantly skirmishing with the hostile cavalry, we took position at Port Republic, nearly one hundred miles south of Winchester, 27 September. On the 28th, Early moved twenty miles further south to drive off two divisions of Torbet's cavalry who had got in our rear and were now destroying the railroad bridge at Waynesboro and the tunnel through the Blue Ridge at Rockfish Gap. Driving a force of cavalry before us, our division (Pegram's) arrived just at night and advancing upon the enemy, drove him off in great haste. On October 1st we marched back down the valley to Mt. Sidney, the main force of the enemy being then at Harrisonburg.

Early having been reinforced by Rosser's cavalry brigade and Kershaw's infantry division from Lee's army, prepared to attack the enemy at Harrisonburg. But in the night of the 5th Sheridan retreated down the valley. Early followed and took position at New Market with his infantry. Rosser's and Lomax's cavalry pressing forward near Fisher's Hill, were encountered by a superior force and driven back in confusion, losing eleven pieces of artillery. In fact some wag suggested that the guns that Lee was sending Early about this time be labeled "General Phil. Sheridan, in care of General Jube Early."

THE BATTLE OF CEDAR CREEK.

The object of the valley campaign was to keep the largest possible Federal force detached from Grant to protect the approaches to Washington, the acumen of Lee telling him that the nervous Washington officials would see that the protecting force was a liberal one. Early learning that Sheridan was about to send troops back to Grant, moved farther down the valley on October 12th. On the 13th we reached Fisher's Hill, part of the forces advancing as far as Hupp's Hill.

Finding Sheridan's position across Cedar Creek too strong for a front attack, Early after having it closely scrutinized from the signal station at Massanutten Mountain, determined to surprise and turn the Federal left flank. We moved out at 9 o'clock on the night of the 18th in great secrecy. Canteens were closely strapped to sides to prevent rattling and only whispering allowed. Crossing the turnpike we went around the mountain's base by a trail that wound around over the swift dashing stream. The moon was full and our long line of bayonets glittered in its beams. Just at daybreak we waded the stream. The shot of a Federal picket rang out. We rushed forward with loud yells right into the sleeping camp. A little later in the morning our division had a hand-to-hand engagement with and drove back a larger part of the Sixth corps and aided by Battle's Alabamians, captured six pieces of artillery, which were most bravely defended, the artillerymen dying at their guns rather than surrender. Our division was then moved to the north of Middleton and took position across the pike. Here it remained during the day skirmishing with the cavalry force in its front.

Meanwhile the tide of battle, so strong in our favor in the morning, finally turned. The Confederate commands had been greatly weakened by men who left the ranks to loot the captured camps, so tempting to ill-fed, ill-equipped soldiers. The routed Federals were halted and reformed. Sheridan, absent in the morning, came up, made new disposition and assailed Early in flank. Then came disaster quick on the heels of disaster.

Our command was in position where we could see the line as it broke, first at the point held by Gordon and then at that held by Ramseur. These divisions retired from the field in great disorder. Johnston's brigade was the only organized body that retreated from the face of the enemy with its line unbroken, halting and firing repeatedly as they were pressed upon. In fact they were then the only organized force in Early's whole army. After falling back near Cedar Creek, General Pegram sent an order to General Johnston "to cross the bridge" and follow the road towards Strasburg. General Johnston sent a message saying that it would be impossible to cross the bridge, as the breastworks built by the enemy commanded the bridge completely, and the enemy would occupy them before he (Johnston) could cross; but that he could cross below and preserve his brigade intact. A second staff officer from General Pegram ordered Johnston to bring his brigade across the bridge just under the command of these works which in the meantime, had been occupied by the enemy. While the brigade was attempting to obey the order and cross the bridge, a hot fire was poured into it from these works. Being totally unprotected and at the mercy of the enemy and their formation broken by the rush of fugitives, the brigade fell into confusion and retreated under cover of the saving darkness.

General Early says that could 500 men have been rallied after the creek was passed the pursuit which was feeble, could have been checked sufficiently to have saved not only his own artillery and trains, but also to have brought off the captured guns, all of which got safely over Cedar Creek, but were captured afterwards. Now in Johnston's brigade he would have had a large part of the necessary 500. In view of this Gen-

120   **NORTH CAROLINA - 23rd REGIMENT INFANTRY - Roster and History**

****

eral Pegram's peremptory order to Johnston to cross at the bridge was exceedingly unfortunate. For Early lost, not only all the captured guns, but likewise every piece of his own artillery. A bridge broke on a very narrow part of the road between Fisher's Hill and Strasburg, the artillery and trains could not cross and being undefended, were taken by a small force of Federal cavalry.

Halting at Fisher's Hill till 3 o'clock the next morning the retreat was continued without halt to New Market, nearly thirty miles distant. On this retreat and while near Mt. Jackson, General Johnston was ordered to face about and hold the enemy in check. He formed line of battle, threw out skirmishers, and had one of the hottest fights in which the brigade was engaged on the skirmish line during the war. The enemy was defeated and driven back.

At New Market we rested undisturbed during the remainder of October Recruits and stragglers came in. Dejected spirits revived. The Confederate soldier was himself again, dogged, indomitable. The order to advance once more down the valley was received with joy. Starting 10 November on the 11th we approached Cedar Creek, our last unfortunate battle ground. Sheridan's main force fell back to Winchester. Driving the cavalry before us we reached Newton, within a few miles of Winchester. Making as great a show of force so as to hold as many of the enemy here and away from Lee as possible, we remained here the 11th and 12th, constant skirmishing going on between the opposing cavalry forces. Being too weak to attack Sheridan and he refusing to leave his intrenchments to attack us, we retreated on the night of the 12th, returning to New Market.

Our brigade formed part of the forces returned to Lee's army about the last of November. Camping near Waynesboro, on the following night, we took cars for Richmond. We arrived in the Confederate Capital amid a hard snow storm. The haste in which we were detrained, double quicked through the streets and entrained for Petersburg told us that Grant was still hammering at the defences and that we were sorely needed. This time he was making another effort to get possession of the Petersburg and Weldon Railroad. We were hurried to a point a few miles south of Petersburg to frustrate his attack.

THE BATTLE OF HATCHER'S RUN.

Then followed much arduous picketing on Hatcher's Run, the winter being a severe one. The Twenty-third took a prominent part in the battle of Hatcher's Run, fought in February. It was in the hottest of the fight near the Crow house. At one time when the opposing lines of battle were less than one hundred yards apart, the flag of the Twenty-third was advanced three times, each time falling as its gallant bearer was shot down. Captain A. D. Peace, in command of the regiment, now took up the flag and rushed forward, followed by the men. But just then came the tidings that Pegram had fallen and that we were flanked, and the lines broke and were falling back in confusion till Gordon dashed to the front, restored the fight and the enemy were driven back.

Our regiment lost heavily in the fight, in proportion to numbers. Captain Frank Bennett, in command of the skirmishers that day, lost an arm. Every year of the war had in store a wound for this gallant officer. The day before Seven Pines, in 1862; Chancellorsville, in 1863; Spottsylvania, in 1864; Hatcher's Run, in 1865, are the dates of his woundings.

General Pegram, our division commander, was killed at Hatcher's Run and General James Walker assumed command of the division. Soon after the battle our brigade was sent back to North Carolina, going into camp at Garysburg, our first point of rendezvous in the hopeful days of 1861. Four years of war had dealt hard with the old Twenty-third. Hardship, disease and Yankee lead had left but a battered remnant of the buoyant band of yore.

Remaining here a few days, we were then put on round duty. There were so many men, mostly conscripts, deserting from Lee's army and passing southward through North Carolina, that the Confederate authorities sought to check it by drawing a cordon of troops across their route. Johnston's brigade was the one selected for this duty. Some of the companies were sent back to their neighborhoods to catch deserters. Company A went back to Richmond County. The battalion of sharpshooters went to cope with the recalcitrant mountaineers and possibly other companies elsewhere. But the most of the brigade formed the cordon on Roanoke river, from Gaston to Clarksville, guarding every road and ferry. Our regiment was assigned to the lower end of the line near Gaston with headquarters at Warren Plains. Here we remained for about a month catching probably as many deserters as we had men—which was not many.

March 23rd Gordon telegraphed Johnston to bring his brigade to Petersburg at once. The return was so sudden that the troops far up the river near Clarksville, did not reach the railroad in time and with the detached companies, in other parts of the State, joined us at Petersburg some days later.

It was known to the troops on that night that the next day we were going back to Lee's stern battle ground around Petersburg. Some of the men, loosing that night the captured deserters, fled with them under cover of darkness. But not many and those few were conscripts, men forced into the ranks.

Bivouacing at Stony Creek the first night and marching around the gap in the railroad made by the enemy we pressed on to the front. The night of the 24th we slept on the hard pavements of Petersburg, the last sleep but one of many a gallant fellow that neither hardships, nor wounds, nor even despair, could part from the Southern standard—nothing but death.

While it was yet dark on Friday morning, 25 March, the men were roused, thrown into column and marched silently and rapidly to the east.

We had been chosen part of the forlorn hope of the desperately straitened Confederacy—honor high, but dangerous. Lee's last hope was by a sudden and desperate assault on Grant's left at Fort Steadman to roll back the hostile line and loosen the strangling folds drawn around the Confederate Capital and its sister city on the Appomattox.

The opaque east grew vaguely translucent. The Federal works on Hare's Hill rose in sharp outline against the brightening back ground. Then Gordon's assaulting force, consisting of our division (Walker's) and Grimes' division, sprung over the Confederate works and rushed forward. Hacking, tearing our way through the hostile abatis, we pressed onward under fire too hurried to be other than wild. In a few minutes Fort Stedman and a large section of the left of Grant's works was in our hands and our part of the line, at least, had penetrated several hundred yards further. But the troops expected to support us failed to appear. For an hour or more we held on. Broad daylight came. Gun after gun, battery after battery, from the right, the left, the rear of the Federal line was brought to bear, till it is said that 200 guns were concentrated on us. No command to retreat reached us, but we could see the other troops being driven back. To remain where we were or to attempt to retreat meant, it seemed, annihilation. About two thousand of the assaulting force surrendered. Our brigade was among the troops that came back. Every foot of the retreat was swept by a tremendous tempest of shot, shell, grape, canister—every missile that the engines of war cast from their iron lips. The artillery ploughed and tore up the ground so ceaselessly that in all but color the flying earth looked like a wind driven snow storm.

The Twenty-third had not many men to lose, but of these few a large proportion fell; how many there are no records to tell. General R. D. Johnston, commanding the brigade, sustained a severe sprain of the ankle as he climbed the Federal works, while unfelt for a few minutes in the excitement of the battle, it soon rendered him unable to walk for the rest of the war. Colonel Lea, of the Fifth Regiment, commanded the brigade for the remaining weeks of the war.

Soon after the bloody and unavailing assault on Fort Steadman, our brigade was moved out and placed on picket between Swift Creek and Appomattox river. Here for a little while we had rest. Early on Sunday morning, 2 April, the brigade leaving its position on picket, was hurried on the double quick through the streets of Petersburg. The enemy had broken over and captured part of the works held by Grimes' division and we were the only available troops to retake it. This break in the line was about a mile south of that part of the line we passed over to carry Fort Steadman.

Captain B. M. Collins, of the Twelfth Regiment, then Adjutant-General of the brigade, gives me the following graphic account of this, the last day's fighting around the doomed city: "We moved out through the covered way, nearly knee deep in mud. We could see our captured works swarming with blue-coats. The fire was so hot that to expose an inch of the person above the protection meant death or wounding. Colonel Lea sent me back to report to General Walker, our division commander, that to assault such a force with his weak brigade of about 250 men was a desperate undertaking.

General Walker repeated the order to assault, adding that Captain Hobson (father of Lieutenant Hobson, the hero of Santiago) commanding a force of sharpshooters, would make a diversion in our favor. The diversion amounted to nothing. We crept up within one hundred yards of the enemy, sprang from the ditch and charged. A small part of the works were taken in this rush. This position we set to work to widen, shooting to right and left along the line. There were traverses along the works at frequent intervals made of timber and earth. The ends of the traverses next to the works were roughly fitted, leaving many holes and openings. Through these holes some of the men fired away at light-blue legs while the bulk of the command fired over the traverses at dark-blue heads.

The Federals fought us, but not with the spirit which their immense superiority in numbers would have justified. An attack half as vigorous as ours must have swept over us and captured Petersburg in an hour. For a while no attack at all came from our front. A part of our command was thrown forward recapturing Fort Mahone in advance of the line of works. This opened the hornets nest on us. An overwhelming force of red pantalooned Yankees, sweeping contemptuously across our weak front, recaptured the fort, our troops escaping, bringing the garrison as prisoners. But the charging force paid dear for their temerity. Our deadly enfilading fire piled the ground with red breeches as their flank came by us.

"About midnight came the order to withdraw, telling us that all was up. Passing through Petersburg we were rejoined by the skirmish line left behind that morning in the sudden movement to recapture the works. We then crossed the canal and retreated up the south bank of the Appomattox river, shells falling around us as we went.

"Our depleted corps formed the rear and wagon guards on the retreat, fighting constantly. The enemy brought up fresh troops when one command was worn out. We were under ceaseless strain. I was afraid to sleep lest I be left behind, taking only short, restless naps when completely exhausted."

The handful left of the 23rd fought at Amelia Court House and at Sailor's Creek. Here the whole corps was broke, but got in fair shape by the next day.

General R. D. Johnston suffering too acutely from his swollen ankle to mount a horse, accompanied the retreat in an ambulance. On one occasion finding that the Federal cavalry was about to capture the whole line of wagons and ambulances, he collected a few stragglers, stopped an ammunition wagon, made every man get down and take a gun and with this motley force prevented the capture of the train.

Further on in the great retreat the hostile cavalry broke into the line and captured General Johnston's ambulance and the rest, including a portion of the wagon trains. General Johnston cut the insignia of rank from his coat, mounted a mule, the rider having fled, rode back, organized a force of stragglers and recaptured the whole line.

At midnight of 8 April, we had a bloody skirmish. Before sun up of the fateful 9th the brigade passed swiftly through the little town of Appomattox. Forming a line to the left of the Lynchburg road we made our last charge against dismounted cavalry in a body of woods. The hostile force was swept back in precipitation.

Then for the last time rang out from our thin line, the "Rebel Yell," which had so long heralded the resistless charge of the men in gray.

But then comes an order to halt and to right-about face. We are marched back towards the village, near which the remnant of the Army of Northern Virginia seems to be concentrating. Strange apparitions greet our eyes. Officers in Federal uniform ride unchallenged among our troops. We rub our eyes as if they did not serve us true. But the officers in blue still come and go.

Slowly, heavily, crushingly the agonizing fact bears down upon our hearts. The thing that could not happen had happened. The end of all things was at hand. *Lee had surrendered.*

It is said that the last man to fall was a member of the First Battalion of North Carolina Sharpshooters, attached to our brigade, and that Captain B. M. Collins, of the Twelfth North Carolina, fired the last musket fired by Lee's army.

The greatest of Greek painters in depicting the mental agony of a hero shows him with his face covered, leaving to the imagination the supreme expression of sorrow. We shall so deal with the emotions that filled our breast. Words are futile things when we would describe feelings like those that weighted the Confederate breast. Better leave to the sympathetic imagination which has followed these men from the beginning—which has seen with what valor, what fortitude, what matchless self-devotion they upheld the cause of Southern Independence, to measure the otherwise fathomless abyss of their sorrow and despair at seeing it stricken down forever.

Dr. R. I. Hicks, now of Warrenton, Va., the faithful and efficient surgeon of the Twenty-third throughout the war, says of the regiment: "It did as much hard service, fought as many battles, was as constant in the performance of duty as any other regiment in the army. And at Appomattox it surrendered about as many men as any other regiment in the army." According to the parole list, Johnston's Brigade then numbered 463 men, rank and file.

The authors are well aware that the foregoing sketch is but a meagre and unworthy history of the command whose deeds and sufferings they would fain chronicle. More than the third of a century has passed since the Twenty-third stacked arms for the last time at Appomattox. On many comrades, depositories of priceless reminiscences, death has set all too soon the seal of silence. Even with the living time is fast blurring the scenes that were long so clear and sharply cut that it seemed they must abide with us forever. But the writers have garnered what little could be saved before it was too late, grieving that the harvest should be so poor.

Many a gallant deed has passed into oblivion with him who performed it and the few who witnessed it. Of some individuals a good deal is recorded, of many, nothing. This must not be taken to mean that the men whose gallant deeds are given are the only worthy or even the most worthy. Largely owing to chance, the memory of some brave acts and of the men who performed them survives; while others, perhaps even more gallant, have been lost. Such authentic ones as could be collected the writers have given, deploring none the less that time should have been so partial in his treatment of these comrades in arms, preserving the deeds of some, casting to oblivion the deeds of others.

Vines E. Turner,

Raleigh, N. C.

H. Clay Wall,

Rockingham, N. C.

Note.—In rewriting Sergeant Wall's sketch of the regiment, it gives me great pleasure to acknowledge the valuable assistance of Mr. O. W. Blacknall (son of the late Colonel Chas. C. Blacknall), who has visited the important battlefields of the Army of Northern Virginia and has given much study to Lee's campaigns.

He also has had access to the private letters and papers of his gallant father which have enabled him to rescue from oblivion many interesting and important facts relating to the history of the regiment.

V. E. Turner.

Raleigh, N. C.,
    9 April, 1901.

D